D1131681

Intelligent Control of Robotic Systems

International Series on
MICROPROCESSOR-BASED AND INTELLIGENT SYSTEMS ENGINEERING

VOLUME 25

Editor

Professor S. G. Tzafestas, *National Technical University of Athens, Greece*

The titles published in this series are listed at the end of this volume.

Intelligent Control of Robotic Systems

by

DUSKO KATIC

Robotics Laboratory,
Mihailo Pupin Institute Belgrade, Serbia,
Serbia & Montenegro

and

MIOMIR VUKOBRATOVIC

Robotics Laboratory,
Mihailo Pupin Institute Belgrade,Serbia,
Serbia & Montenegro

KLUWER ACADEMIC PUBLISHERS

DORDRECHT / BOSTON / LONDON

A C.I.P. Catalogue record for this book is available from the Library of Congress.

ISBN 1-4020-1630-1

Published by Kluwer Academic Publishers,
P.O. Box 17, 3300 AA Dordrecht, The Netherlands.

Sold and distributed in North, Central and South America
by Kluwer Academic Publishers,
101 Philip Drive, Norwell, MA 02061, U.S.A.

In all other countries, sold and distributed
by Kluwer Academic Publishers,
P.O. Box 322, 3300 AH Dordrecht, The Netherlands.

Printed on acid-free paper

Printed in the Netherlands.

Contents

List of Figures

List of Tables

Preface

As robotic systems make their way into standard practice, they have opened the door to a wide spectrum of complex applications. Such applications usually demand that the robots be highly intelligent. Future robots are likely to have greater sensory capabilities, more intelligence, higher levels of manual dexterity, and adequate mobility, compared to humans. In order to ensure high-quality control and performance in robotics, new intelligent control techniques must be developed, which are capable of coping with task complexity, multi-objective decision making, large volumes of perception data and substantial amounts of heuristic information. Hence, the pursuit of intelligent autonomous robotic systems has been a topic of much fascinating research in recent years.

On the other hand, as emerging technologies, Soft Computing paradigms consisting of complementary elements of Fuzzy Logic, Neural Computing and Evolutionary Computation are viewed as the most promising methods towards intelligent robotic systems. Due to their strong learning and cognitive ability and good tolerance of uncertainty and imprecision, Soft Computing techniques have found wide application in the area of intelligent control of robotic systems.

Our objective in writing the present book was to produce a fair combination of comprehensive report, survey, theoretical background and special research works with appropriate application examples in the area of intelligent control of robotic systems. This book is chiefly focused on theoretical and application aspects of neural networks, fuzzy logic, genetic algorithms and hybrid intelligent techniques in robotics. The specific emphasis in research work is given to the development of efficient learning rules for robotic connectionist training and synthesis of neural learning algorithms for low-level control in the domain of robotic compliance tasks. The book contains several different examples of applications based on neural and hybrid intelligent techniques.

The book is organized in the following way. The first chapter gives an introduction to intelligent control together with a presentation of the basic ideas of intelligent control in robotics. Special attention is paid to the role of learning as one of the main intelligent capabilities of the control algorithms dealt with in this book.

The focus then shifts (Chapter 2) to neural networks (connectionist systems), to review the fundamental concepts and learning principles, discus important implementation issues, and provide a survey of connectionist algorithms in contemporary robotics. The connectionist models for robotic purposes are the subject of a special analysis. The issues of neural networks in robotics with the applications in the domain of kinematic, dynamic, and sensor-based learning are analyzed. A special part of this chapter represents the development of new, efficient learning structures for robot training, followed by appropriate simulation examples. The various implementation issues are addressed by design exploration and the verification of intelligent control paradigms for a variety of robotic applications, including industrial, service, mobile, locomotion, space, underwater and other types of robotics.

In the following chapters, the fundamentals of fuzzy logic (Chapter 3), genetic algorithms (Chapter 4) and hybrid intelligent technics (Chapter 5) are presented together with a comprehensive report on their importance and recent applications in autonomous robotic systems.

The synthesis of new, advanced learning algorithms for robotic contact tasks by nonrecurrent and recurrent connectionist structures is presented in Chapters 6 and 7 as the main research contribution. In Chapter 6, which includes a survey of connectionist algorithms for robotic contact tasks, the main concern is the development of learning control algorithms as an upgrade of conventional non-learning control laws for robotic compliance tasks (algorithms for stabilization of robot motion, stabilization of robot interaction force and impedance algorithms). In view of the important influence of robot environment, a new, comprehensive learning approach, based on simultaneous classification of robot environment and learning of robot uncertainties, is reported in Chapter 7. The proposed comprehensive algorithms include the synthesis and application of two newly proposed classifiers: pure perceptron classifier and wavelet network classifier. Both chapters are accompanied by simulation studies to validate the proposed algorithms.

The book concludes with Chapter 8, which presentates some interesting examples of connectionist approaches, together with some supporting intelligent techniques for special robotic applications. The examples include connectionist reactive control of the soft-sensored grippers robotic assembly tasks, a special genetic-connectionist algorithm for compliant robotic tasks and a connectionist approach to robotized road vehicles (automobiles).

This book has evolved from many years of research work and teaching in the areas of automatic control, intelligent robotics and robotics in general. There is one unifying aspect to the work reported in the book, namely its interdisciplinary nature, especially in the combination of robotics, control theory, computational intelligence and soft computing paradigms. It is our hope that it will be useful to a wide audience of engineers, ranging from students and academic researchers, to the practitioners. The presented text can satisfactorily serve as an educational tool for engineering students interested in pursuing the study of intelligent autonomous robotic systems, as well as a starting base for researchers in the ongoing research in these areas.

Belgrade, April 2003

DUSKO KATIC, MIOMIR VUKOBRATOVIC

Acknowledgments

During the years taken to research and write this book, consciously and subconsciously, we have picked up material from a knowledge base called intelligent control in robotics. Hence, we have had the pleasure and privilege of interacting with many researchers throughout the world and to whom we owe our deepest thanks. The long-term support of our research on intelligent control in robotics by the Serbian Ministry of Science and Technology is gratefully acknowledged.

We would like to express our great appreciation to Branislav Borovac for our collaborative work on the problems of connectionist reactive control for robotic assembly tasks (Chapter 8). We want to thank Aleksandar Rodic for our collaborative research in the area of neural controllers for robotized vehicles (Chapter 8).

We also want to express our gratitude to Branko Karan for his participation in writing some parts of the text about the general theory of fuzzy logic and their application to robotic systems (Chapter 3).

Finally, we thank Luka Bjelica who has proofread the final version of the manuscript and thereby removed some of the errors and gave us valuable suggestions for improvements.

Chapter 1

INTELLIGENT CONTROL IN CONTEMPORARY ROBOTICS

1. Introduction

Modern technological systems are characterized by poor system and subsystem models, high dimensionality of the decision space, distributed sensors and decision makers, high noise levels, multiple subsystems, levels, time-scales and/or performance criteria, complex information patterns, overwhelming amount of data and stringent performance requirements. Hence, contemporary research in technological systems is oriented towards multi-disciplinary studies based on the synthesis and application of various control and management paradigms needed for efficient realization of the complex technological system goals and requirements and coping at the same time with all the mentioned problems and constraints. In the scope of these facts, modern engineering approaches have begun to emerge in designing, constructing, testing, and exploiting practical technological systems with the capabilities of:

a) Intelligent production planning, resource management, and task scheduling,

b) Intelligent decision making, task decomposition, goal pursuit, and reaction to unanticipated situations,

c) Intelligent information organization, processing, perception and situation understanding,

d) Intelligent path planning for automated route selection, navigation and obstacle avoidance,

e) Intelligent sensing of the environment and internal state of the system,

f) Intelligent learning from experience and instructions,

g) Intelligent control of motion precision, speed, position and force actuation.

A common characteristic of all the above system's capabilities is the use of intelligent features as essential properties. Intelligence as an essential mental quality can be defined in various ways, but one possible definition [5] is: *"Intelligence* is an ability of a system to act appropriately in an uncertain environment, where appropriate action is that which increases the probability of success, and success is the achievement of behavioural subgoals that support the system's ultimate goal". Another important definition of intelligence is the ability to reach a wide variety of the system's goals. The more goals a system can handle, the more intelligent the system. Based on intelligence properties, it becomes possible to build intelligent control systems that can both collect and process information, as well as generate and control behaviour in real time, so as to cope with the situations that evolve amid the complexities of natural and man-made competition in the real world. Hence, intelligent control systems in the frame of a technological system are the control systems with the ultimate degree of autonomy in terms of self-learning, self-reconfigurability, reasoning, planning and decision making and the ability to extract the most valuable information from unstructured and noisy data from any dynamically complex system and/or environment.

However, the question is: What are the reasons for using an intelligent control paradigm instead of conventional control and management paradigms?

There are many reasons. For instance, the difficulties that arise in the control of complex technological systems can be broadly classified into three categories. The first is the complexity of the system's requirements and goals, the second is the presence of nonlinearities and constraints, and the third is the increased uncertainty in the system modeling and designing. Changes in the environment and performance criteria, unmeasurable disturbances and component failures are some of the characteristics which necessitate intelligent control.

On the other hand, in the design of controllers for complex dynamic systems there are needs that cannot be successfully met by the existing conventional control theory. In conventional control, the goals of a controller are fixed and defined by the designer. While this may be adequate for controlling most plants, it is insufficient for control of real-life situations that require some form of autonomy . These autonomous systems must set goals on their own, depending on recognition of the situation (perception), and utilization of background knowledge (experience) and with further constraints and goals imposed by higher levels of the system. Such systems may require reasoning under uncertainty, which often involves a great deal of judgmental imprecision. Furthermore, the control system may need to make decisions in an unstructured environment, these functions being performed in the past by human operators of mundane tasks. To protect the system from hazards, a high degree of autonomy is desired. Hence, high-level decision-making techniques for reasoning and taking actions under uncertainty and taking actions must be utilized. Intelligent control methodologies enhance and extend traditional control methods. Intelligent controllers are envisioned, that emulate human mental faculties such as adaptation and learning, planning under great uncertainty and coping with large amounts of data in order to effectively control complex processes. Intelligent control systems may be autonomous, human-aided, or computer-aided. They have the potential to yield new levels of system performance in terms of reliability and accommodation of significant changes in the system parameters.

It is clear that conventional control techniques have not been able to provide complete autonomy or semi-autonomy for such systems. The main reason is that the current techniques do not carry the *human-like* intelligence that provides ability to learn by generalization and make decisions in the unstructured environments. Thus, we need intelligent control systems that would be able to distribute the decision-making load among a team of decision - makers by decomposing functions into specific tasks and assigning these tasks to individual agents, who are responsible for their performance. This is a hierarchical structure.

Intelligent autonomous control systems may be composed of either one or multiple self-contained entities - often called agents - each being able to perform autonomously. Those agents interacting with an immediate environment, are expected to display both *reactive behaviour* and *goal directed behaviour*. The latter is related to the agent's ability to behave in line with long-term goals, which are usually specified in advance, while reactive behaviour is driven by the agent's immediate response to changes in its environment. It may happen that reactive behaviour and goal - directed behaviour are in mutual conflict, and consequently the agent's reasoning ability is also required to resolve the conflict. Hence, an important approach is intelligent robotic agent control, acting in a real-world environment based on a lifelong learning approach combining cognitive and reactive capabilities [147].

When an intelligent autonomous system is composed of multiple agents, then, in order to construct a system, they must also display *cooperative behaviour*, which is composed of : (i) the ability to identify the existence of the other agents (particularly those that are part of the same system); (ii) the ability to coordinate their actions so as not to interfere with each other; (iii) the ability to predict the course of action of other agents; and (iv) the ability to share the system goal(s). Thus, cooperation behaviour is much more than just the coordination and synchronization of the activities of individual agents and appears to be the focal attribute of intelligent autonomous agents, particularly of those aimed at participating in teamwork and sharing system goals. For the same reasons, the agent's ability to cooperate is considered to be the foundation of the study of intelligent autonomous control systems.

Intelligent control as a new discipline introduced by K.S.Fu [82],[242] has emerged within the cluster of classical control disciplines with primary research interest in the specific kind of technological systems (systems with recognition in the loop, systems with elements of learning and self-organization, systems which sometimes do not allow for representation in a conventional form of differential and integral calculus). A primary concern of intelligent control is the cross-disciplinary studies of high-level control in which control strategies are generated using human intelligent functions as perception, simultaneous utilization of memory, association, reasoning, learning or multi-level decision making in response to fuzzy or qualitative commands. One of the main objectives of intelligent control is to design a system with acceptable performance characteristics over a very wide range of structured and unstructured uncertainties.

In view of the high and complex demands they have to meet as important integral parts of flexible manufacturing systems, robots are ideal objects for the application of intelligent control. Intelligent robots are functionally oriented devices built to perform sets of tasks instead of humans, as autonomous systems capable of extracting information from environment using knowledge about world and intelligence of their duties and proper governing capabilities. Intelligent robots should be autonomous to move safely in a meaningful and purposive manner, i.e. to accept high-level descriptions of tasks (specifying what the user wants to be done, rather than how to do it) and should execute them without further human intervention. Also, intelligent robots are autonomous robots with versatile intelligent capabilities, which can perform various anthropomorphic tasks in a familiar or unfamiliar (structured or unstructured) working environment. They have to be intelligent to determine all the possible actions in an unpredictable dynamic environment using information from various sensors. Human operator can transfer knowledge, experience and skill in advance to the robot to make it capable of solving complex tasks. However, in the case when the robot performs in an unknown environment, the knowledge may not be sufficient. Hence, the robot has to adapt to the environment to be able to acquire new knowledge through the process of learning.

The first approach to making robots more intelligent was the integration of sophisticated sensor systems such as computer vision, tactile sensing, ultrasonic and sonar sensors, laser scanners and other smart sensors. However, in order to ensure high-quality control in robotics, new intelligent control techniques have to cope with task complexity, multi-objective decision making, large volume of perception data and substantial amounts of heuristic information. Moreover, there are many sources of undesired behaviour of robotic systems when working according to conventional control algorithms that can be eliminated by novel intelligent algorithms. For example, it is well known that conventional control algorithms are model-based schemes that in most cases are synthesized using incomplete information and partially known or inaccurately defined parameters. Conventional control algorithms are also extremely sensitive to the lack of sensor information or fuzzy sensory information and to unplanned events and unfamiliar situations in the robot working environment. The robot is not able to capture and use past experience and available human expertise. All previously mentioned facts and examples provide a motivation for robotic intelligent control, which is capable of ensuring that the manipulation robot can sense its environment to process the information necessary for uncertainty reduction, to plan, generate and execute high-quality control action. Hence, efficient robotic intelligent control systems must be based on the following features:

a) autonomy in decision making for all hierarchical control levels,

b) robustness and great adaptability to system uncertainties and environment changes,

c) learning and self-organizing capabilities with generalization of the acquired knowledge,

d) skill acquisition based on acquisition of expertise and experience,

e) implementation in real time using fast processing architectures for sensor fusion and control computation.

The learning properties of intelligent control algorithms in robotics are very important as they ensure the achievement of high-quality robot performance. Our knowledge of robotic systems is in many cases incomplete, because it is not possible to describe the system's behaviour in a rigorous mathematical way. Hence, use is made of the learning of active compensation for uncertainties, which results in continuous improvement of robotic performance. It is well known that conventional adaptive and non-adaptive robot control algorithms comprise the problem of robot control during execution of particular robot trajectories without considering repetitive motion. Hence, in terms of learning, almost all manipulation robots have no memory. Thus, previously acquired experience about the dynamic robot model and control algorithms is not applied in the robot control synthesis. It is expected that the use of a training process by repeating the control task and recording the results accumulated in the entire process will steadily improve performance. The state variables-dependency of robot dynamics may also be solved by learning and storing solution, while time-dependency of robot parameters requires an on-line learning approach. If the learning control algorithm once learns a particular movement, it will be able to control quite a different and faster movement using the generalization properties of the learning algorithm.

There are several intelligent paradigms that are capable of solving intelligent control problems in robotics. Beside symbolic knowledge-based systems (AI - expert systems), connectionist theory (NN - neural networks), fuzzy logic (FL), and theory of evolutionary computation (GA - genetic algorithms), are of great importance in the development of intelligent robot control algorithms. Also of great importance in the development of efficient algorithms are hybrid techniques based on integration of particular techniques such as neuro-fuzzy networks, neuro-genetic algorithms and fuzzy-genetic algorithms.

Intelligent control systems can benefit from the advances in artificial neural networks [285], [115] as a tool for on-line learning optimization, and optimal policy making. The connectionist systems (neural networks) represent massively parallel distributed networks with the ability to serve in advanced robot control loops as learning and compensation elements using the abilities of nonlinear mapping, learning, parallel processing, self-organization and generalization.

The fuzzy control systems [410], [341] based on mathematical formulation of fuzzy logic have the characteristic to represent human knowledge or experience as a set of fuzzy rules. Fuzzy robot controllers use human know-how or heuristic rules in the form of linguistic if-then rules, while a fuzzy inference engine computes the efficient control action for a given purpose.

The theory of evolutionary computation with genetic algorithms [97],[114] represents an approach to global search optimization, which is based on the mechanisms of natural selection and natural genetics. It combines survival of the fittest among string structures with a structured yet randomized information exchange to form a search algorithm with expected ever-improving performance.

Each of the proposed paradigms has its own merits and drawbacks. To overcome their drawbacks, appropriate integration and synthesis of hybrid techniques are needed for efficient application in robotics. Symbiotic intelligence incorporates a new type of robotic system having many degrees of freedom (DOFs) and multi-modal sensory inputs. The underlying idea is that the richness of inputs to and outputs from the system, along with co-evolving complexity of the environment mixed with various intelligent control paradigms, is the key to the emergence of intelligence. For example, neuro-fuzzy networks represent a combined tool by which a human operator can import his knowledge by means of membership functions. On the other hand, membership functions are modified through the learning process in the same way as fine tuning by neural networks. After learning, the human operator can understand the acquired rules in the network. Neuro-fuzzy networks are faster than conventional neural networks in terms of convergence of learning. On the other hand, fuzzy logic and neural networks can serve as evaluation functions for genetic algorithms. At the same time, genetic algorithms can be structure optimizers for fuzzy and neural algorithms. Computational intelligence techniques map well onto nonlinear problems and are better at handling uncertainties.

All these techniques may be incorporated in advanced and sophisticated robotic control systems that were in general inspired by biological designs and neurobiological principles [93], [90], [186]. In the last two decades, many researchers explored the design of autonomous systems, swarms of intelligent agents and biologically inspired control designs and actuators [68], [103], [354], [32], [238], [173], [11], [37], [172]. The research in this area is specially oriented toward the ideas of artificial life and adaptive behaviour. An ultimate goal is the creation of a humanoid robot as an autonomous agent, which is capable of mimicking all aspects of human action, perception and cognition in everyday life and in remote and unfriendly environments.

The mentioned intelligent control paradigms are inherently connected to the hierarchical structure of the intelligent robot control system [293], [351], [304]. One of the proposed solutions for the hierarchical structure of an intelligent robot control system is shown in Figure 1.1. The proposed hierarchical robot intelligent control system consists of three levels: adaptation level, skill level and learning level. There are also three main feedback loops. The learning level is based on an expert system for inference on two hierarchical levels: recognition and planning to develop control strategies. On the recognition level, neural networks or fuzzy neural networks can transform various pieces of numeric information from sensors into symbolic quantities, and perform sensor fusion and production of meta-knowledge for learning. On the planning level, the system reasons symbolically for strategic plans and schedules of robot motion such as task, path, trajectory, position, force and other planning in conjunction with the knowledge bases. The aim of the learning level is the production of control strategies for the skill level and adaptation level. Knowledge on the learning level is given by the human operator in the top-down manner and acquired by the heuristics of the skill level and adaptation level in the bottom-up manner. On the skill level, the aim of the neuro-fuzzy network is to generate the appropriate control reference signals for control strategies formed by the learning level. On the adaptation level, connectionist systems are used to adjust the control law to the current status of the robot and its environment. Particularly, neural networks are oriented to compensate for nonlinearities and uncertainties contained in the robot and its environment.

However, hierarchical intelligent control systems are well suited for structured and highly-predictable manufacturing robotics. These control systems have some drawbacks associated with a perceived lack of responsiveness in unstructured and uncertain environments. On the other hand, reactive intelligent control systems in robotics [11] represent a technique for tightly coupling perception and action, typically in the context of motor behaviour, to produce a timely robotic response in a dynamic and unstructured environment. The reactive systems represent a scientific basis for considering a special research area in intelligent robotic systems, called *behaviour-based robotics* [11]. In these systems, behaviour represent a simple sensorimotor pair, with the sensor system intended for acquisition of the information needed for low-level motor reflex response.

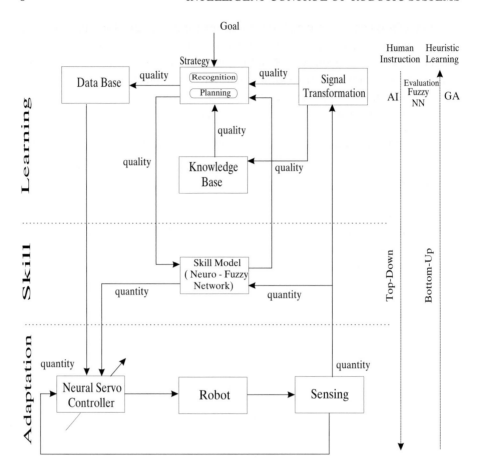

Figure 1.1. Hierarchical intelligent robotic systems

Intelligent control has a very important role in contemporary and future robotic research as it is defined in EU FP6 FET initiative "Beyond Robotics" [75], where inclusion of intelligent paradigms is related to the following research topics in robotics:

1 Cognitive Companions - The development of cognitive robots whose "purpose in life" would be to serve humans as assistants or "companions". Such robots would be able to learn new skills and tasks in an active and non predetermined way and to grow in constant interaction and co-operation with humans. A robotic "companion" can be considered as an adaptive servant that co-exists and continuously interacts with the user. For this purpose, it must evolve with the user so as to acquire the necessary skills, representations and competencies. As an example, it may be used as an "assistant"

that is living and growing together with its master, is adapting itself to and is complementing the mental and physical faculties of its master and is eventually able to compensate for any degradation of its master's performance. For companionship, the robot will naturally be exposed to a wide variety of situations and events. Hence, the system must be able to acquire skills and perform new tasks. In addition, the constant exposure to new environments and settings imposes significant constraints on perception and reasoning capabilities to ensure robust real-time performance over years of use.

2 Human Augmentation - Hybrid bionic systems that would augment human capabilities such as perception of the environment, motion, interaction with other humans, etc. This would involve smooth integration of sophisticated robotic and information systems with human perception-action systems using bi-directional interfaces with the human nervous system. For this objective, systems are to be developed in which there is a tight coupling between the human and the artefacts, e.g., intelligent prosthetics (artificial sensory organs, arms, limbs, etc). The system must be flexible enough in terms of physical interaction and skill/task adaptation to function as a regular member of the human body. High-fidelity, two-way brain interfaces or other more advanced interfaces could provide not only full support to accommodate human functionalities like manipulation or walking, but also for augmenting humans with new perception-action capabilities to go beyond our current physical limitations.

3 Robot Ecologies - The development of autonomous microrobot groups consisting of many heterogeneous members exhibiting collective behaviour and intelligence. The robots would be able to self - organize, adapt, co - operate and evolve in order to attain a global objective. Microrobot groups can be viewed as parallel systems in terms of perception, reasoning and action generation that open up interesting issues for their co - ordination, adaptation and evolution in order to jointly accomplish a wide variety of different tasks. Given a large number of "rational physical agents", the issues of task distribution, co-ordination and sharing of knowledge are as yet unsolved, in particular for heterogeneous systems that have to operate and co-operate in real-world, open-ended environments.

2. Role and Application of Learning in Robotics
2.1 Introduction to robot learning

Robot learning refers to the process of acquiring a sensory-motor control strategy for a particular movement task through a training process by trial and error. The goal of learning control can be defined as the need to acquire a task - dependent control algorithm that maps the state variables of the system and its environment in an appropriate control signal (action). The linear or nonlinear control variable depends also on the system parameters that need to be adjusted by the learning system. The aim of robot learning control is to find a function that is adequate for a given desired behaviour and robotic system.

Generally, robot learning can be considered through three main sub-branches: supervised, unsupervised and reinforcement learning. Supervised learning is needed to form nonlinear coordinate transformations and internal models of the robot environment. For autonomous movement systems, supervised learning has to proceed incrementally and in real time. It is possible to learn the control algorithm directly, where desired behaviour needs to be expressed as an optimization criterion that must be optimized over a certain temporal horizon, resulting in an appropriate cost function. In the cost function, a discount factor is used, that reduces the influence of the criterion in the far future. Such optimization problems have been solved in the framework of dynamic programming [70] and its recent derivative, reinforcement learning [325]. Reinforcement learning is used to allow the robot system to learn from scalar evaluation (*punishment or reward*) only, instead of precise teacher's information (supervised learning). For reinforcement learning, the optimization criterion corresponds to the 'immediate reward', while the cost function is called the 'long-term reward'. The control algorithm is acquired with reinforcement learning by first learning the optimal cost function for every state variable, usually by a technique called temporal difference learning, and then deducing the control algorithm as the control variable in every state variable that leads to the best future performance. Many variations of reinforcement learning exist, including methods that avoid estimating the optimization cost function. However, the main disadvantages of reinforcement learning are a large amount of exploration of all actions and states for proper convergence of learning and the lack of generalization among continuous variables.

A possible way to reduce the computational complexity of the learning of a control algorithm comes from modularizing the control algorithm. Instead of learning the entire control algorithm in one big representation, one could try to learn sub-algorithms that have reduced the complexity and, subsequently, build the complete algorithm out of such sub-algorithms. This approach is also appealing from the viewpoint of learning multiple tasks: some of the sub-algorithms may be re-used in another task such that learning new tasks should be strongly facilitated. Robot learning with such modular control systems, however, is still in its preliminary phase. [326], [256], [297].

For many robotic systems, however a direct control algorithm is not advantageous since it fails to reuse modules in the algorithm that are common across multiple tasks. This view suggests that, in addition to a modularization of robotic motor control and learning in the form of a compound of simpler algorithms, modularization can also be achieved in terms of a functional structuring within each control algorithm. A typical example is to organize the robot control algorithm into several processing stages: strategic, tactical and executive level. While global task planing is realized on the strategic control level, the tactical control level generates a desired kinematic trajectory, i.e., a prescribed way of how the state of the movement system is supposed to change in order to achieve the task goal. The execution control level transforms the desired trajectory into appropriate motor commands. Separating global task and trajectory planning and execution is highly advantageous. For instance, an indirect control approach only requires the learning of the robot execution control level, usually in the form of an inverse model.

From the view point of robot learning, functional modularity also decomposes the learning problem into several independent learning problems. The executive control level can be learnt with supervised learning techniques. For complex movements, trajectory planning requires more sophisticated reinforcement learning methods of direct control, the difference being that control variables are replaced with a desired change in the trajectory. Applying reinforcement learning to kinematic planning is less complex than solving the complete direct control problem, but it remains an open research problem for high dimensional movement systems.

A topic in robot learning that has recently received increasing attention is that of imitation learning. The idea of imitation learning is intuitively simple: a student watches the performance of a teacher, and, subsequently, uses the demonstrated movement as a seed to start his/her own movement. Imitation involves the interaction of perception, memory, and motor control, subsystems that typically utilize very different representations, and must interact to produce and learn novel behaviour patterns. The ability to learn by imitation has a profound impact on how quickly new skills can be acquired [295]. From the viewpoint of learning theory, imitation can be conceived as a method to bias learning towards a particular solution, i.e., that of the teacher. However, not every representation for motor learning is equally suited to be biased by imitation [294]. Imitation learning thus imposes interesting constraints on the structure of a learning system for motor learning. Imitation learning forces us to deal with perceptual and action uncertainty, non-stationarity, and real-time constraints. As a result, imitation approaches to learning, inspired by learning in biology, strive for efficient solutions that can cope with the challenges of real-world domains.

Whether direct or indirect control is employed in a motor learning problem, the core of the learning system usually requires methods of function approximation in the neural network and statistical learning literature. Function approximation is concerned with approximating a nonlinear function (model of the robotic system with its environment) from noisy data.

Many different methods of function approximation exist in the literature. These methods can be classified into two categories, local and global algorithms. The power of learning in neural networks comes from the nonlinear activation functions that are employed in the hidden units of the neural network. Global algorithms use nonlinear activation functions (sigmoid function) that are not limited to a finite domain in the input space of the function. In contrast, local algorithms make use of nonlinear activation functions that differ from zero only in a restricted input domain (Gaussian function).

Despite both local and global algorithms being theoretically capable of approximating arbitrarily complex nonlinear functions, the learning speed, convergence and applicability to high-dimensional learning problems differ significantly [296].

Global learning algorithms can work quite well for problems with many input dimensions, since their non-local activation function can span even huge spaces quite efficiently. However, global algorithms usually require very careful training procedures such that the hidden units learn how to stretch appropriately into all directions. Along with the problem how to select the right number of hidden units, it becomes quite complicated to train global algorithms for high dimensional robot learning problems.

Local learning algorithms approximate the complex regression surface with the help of small local patches, for instance locally constant or locally linear functions [18]. The problem of local algorithms is the exponential explosion of the number of patches that are needed in high dimensional input spaces. The only way to avoid this problem is to make the patches larger, but then the quality of function approximation becomes unacceptably inaccurate. There is theoretically no way out of the curse of dimensionality - but empirically, it turns out not to be a problem. Local learning can exploit this property by using the techniques of local dimensionality reduction and can thus learn efficiently, even in very high dimensional spaces [353].

Applying function approximation to the problems of robot learning requires a few more considerations. The easiest applications are those of straightforward supervised learning, i.e., where a teacher signal is directly available for every training point (state variables, output variables and control variables can be taken directly from sensors). Learning becomes more challenging when instead of the teacher signal only an error signal is provided, and the error signal is just an approximation. Thus, learning proceeds with "moving targets", which is called a nonstationary learning problem. For such learning tasks, neural networks need to have an appropriate amount of plasticity in order to keep on changing until the targets become correct. On the other hand, it is also important that the network converges at some point and averages out the noise in the data, i.e., the network does not have too much plasticity. Finding appropriate neural networks that have the right amount of plasticity-stability tradeoff is a non-trivial problem, and so far, there are many heuristic solutions . Nonstationary learning problems are quite common in robotic learning. Learning the cost function in reinforcement learning (temporal difference algorithm) [325] can only provide approximate errors. Other examples include feedback error learning and learning with distal teachers [155]. Feedback error learning creates an approximate motor command error by using the output of a linear feedback controller as the error signal. Learning with distal teachers essentially accomplishes the same goal, except that it employs a model taught in advance to map an error in sensory space to an approximate motor error.

Robot learning is a surprisingly complex problem. It needs to address how to learn from (possibly delayed) rewards, how to deal with very high-dimensional learning problems, how to use efficient function approximators and how to embed all the elements in a control system with real-time performance. A further difference from many other learning tasks is that in robotic learning the training data are generated by the movement system itself. Efficient data generation, i.e., exploration of the world, will result in fast learning, while inefficient exploration can even prevent successful learning all together. Given the fact that only very few robots in the world are equipped with learning capabilities, it becomes obvious that research on robot learning is still in an early stage of development.

2.2   Application of robot learning

In contemporary technological systems, robotics and automation technology have an important role in a variety of manufacturing tasks. These manufacturing operations can be categorized in two classes, related to the nature of the interaction between the robot and its environment. The first one is concerned with non-contact, i.e. unconstrained motion in a free work space. In non-contact tasks the robot's own dynamics has a crucial influence upon its performance. A limited number of simple and most frequently performed robotic tasks in practice, such as pick-and-place, spray painting, glueing or welding, belong to this group. In contrast to these tasks, many complex advanced robotic applications such as assembly and machining operations (deburring, grinding, polishing, etc.) require the manipulator be mechanically coupled to the other objects. These tasks are denoted as essentially contact tasks, because they include the phases where the robot end-effector must come into contact with objects in its environment, produce certain forces upon them, and/or move along their surfaces. Inherently, each manipulation task involves tcontact with the object being manipulated. The terms *constrained* or *compliant-motion* are usually applied to the contact tasks.

The objective of using adaptive and learning capabilities for the above-mentioned operations is to simplify the implementation process, to improve the system's reliability and thus achieve a tremendous practical impact. Another important characteristic of learning techniques may be the enhanced capability to design robust hierarchical robotic systems. Namely, there are possibilities to use learning control techniques on all hierarchical control levels. Learning behaviour on one hierarchical level will facilitate the technological demands for specific coordination with the other levels. Robotic learning is the ability of a robot to adjust to its dynamic environment and changing task conditions. It can increase flexibility by enabling the robot to deal with different and unexpected situations.

On the strategic and tactical (learning) hierarchical control levels, there are many opportunities for the application of learning systems to task-planning and task-reasoning problems, particularly those that confront the issue of uncertainty in the task environment. Also, a clear opportunity for the impact of learning systems is in the use of sensing and inspection technology for industrial applications. It is important to notice that for these control levels, specific methods from machine learning [42] [11] (inductive learning, case-based learning, explanation-based learning, evolutionary learning, reinforcement learning) are the dominant learning tools for problems on these levels.

Early research in robotic learning began with one robot and one learning entity. Recent improvements in computer speed and cost have made multi-robot systems a promising research topic for robotic learning. A key feature of multirobot systems is the potential to cooperate: several robots can help each other to accomplish a task faster or better, and they can compensate for each other's weaknesses. Early research in multirobot learning began with artificial intelligence concepts and with no physical implementation. In the last decade, there have been many real-robot learning experiments. Reinforcement learning [325], [158] has become a successful learning method for real robotic systems [239], [21], [158]. One of the popular reinforcement learning algorithms is Q-learning [378]. There have been a lot of successful uses of Q-learning on a single robot. However, to date there have only been a small number of learning experiments with multiple robots. Mataric [237] used reinforcement learning to control the behaviour of a robot group. Balch [21] performed an experiment with different types of tasks to explore the effect of diversity in robot groups. Both researchers used modified rewards (i.e., shaped reinforcement signals or progress estimators) to give feedback on progress or specific behaviour. However, there were variations in performance that depended on several factors.

Recently, a number of mobile robots that can demonstrate highly complicated motions have been realized. However, most of these technological products have been strictly designed by skilled engineers who knew well the environment in which the robots would be situated. Therefore, it can be considered that these robots exhibit brittleness against unexpected environmental changes. Because of this, it is expected to endow robots with on-line adaptation and learning abilities. Hence, learning issues under the framework of behaviour-based mobile robots represent an important contemporary research field [191], [373], [327], [20], [236], [262].

The robot with on-line learning ability is expected to demonstrate robustness under real environments, since it can autonomously arrange its sensorimotor map according to the given environment. The integration of autonomous mobile robots in real environments requires conceiving of a robot control strategy (behaviour) capable of dealing with incomplete, imprecise and uncertain robot's perceptions, as well as partially unpredictable actions. Robots will have to be autonomous in unmodelled dynamic environments while behaving in ways useful to humans. The question is therefore how can robots acquire these behaviours? Some behaviours can be explicitly programmed, but this requires an explicit description of the tasks and a model of the environment. Some behaviours can be learned using methods such as reinforcement learning, or genetic algorithms. This again requires, the need to explicitly define the behaviours by the intermediary of a reward or fitness function and the use a trial-and error scheme that is impossible to achieve in most environments.

From the human user point of view, a good way to define a behaviour is to interact directly with the robot in the destination environment. A set of methods that could be grouped under the name of *Empirical Learning* are gaining interest in the literature. These methods are *Learning by demonstrations or from examples*, *Imitation* or more classically *Supervised Learning*.

Our main concern will be the integration of learning techniques for the control of manufacturing operations on the executive (skill and adaptation) hierarchical level. As is well known, a common and simple way for the non-learning control of manipulation robots on the lowest hierarchical level is the use of local PID (proportional-integral-derivative) regulators for each DOF of the robotic mechanism for position tracking and PID regulators for force regulation. However, this control law is not adequate for advanced industrial robots that must demonstrate the capability of high precision and speed in a complex working environment. The influence of couplings between the subsystems and interaction with the environment is substantial, so that we should include a "dynamic" control [361],[356], which uses the dynamic model of the robot mechanism and the model of the robot environment in the process of control synthesis. However, a common problem, especially in manufacturing operations, is how to describe correctly the robot-environment dynamics and how to synthesize control laws that simultaneously stabilize both the desired position and interactive force. For example, in robot contact operations, such as grinding, deburring and polishing, it is essential to control the tangential velocity of the tool along the workpiece and the force normal to the work surface. It is a very difficult task to achieve an acceptable system performance because of a high level of unpredictable interactions between the robot and the environment. Also, various system uncertainties must be taken into account when considering system behaviour. For example, in the procedure of controller design, we have to cope with structured uncertainties (inaccuracies of the robot mechanism parameters, imprecise position of the workpiece, varying degrees of stiffness of the environment, robot tool and robot itself, etc.), unstructured uncertainties (unmodelled high-frequency dynamics as structural resonant modes, actuator dynamics and sampling effects) and measurement noise. Moreover, the time-varying nature of the robot parameters and variability of the robot tasks can cause a high level of interaction force or the loss of contact. In this case, the conventional non-adaptive algorithms are not robust enough, because they can compensate only for a small part of the above-mentioned uncertainties. Hence, a more suitable approach would be the one using adaptive control techniques [361]. The adaptive control technique in robotics was applied as a parameter adaptation technique with the possibility of adaptation in feedforward or feedback loops. All such methods usually comprise on-line schemes with recursive least-squares (RLS) optimization criterion and short-term learning in which past experience is completely forgotten. To summarize, the conventional adaptive control techniques in robotics can tolerate wider ranges of uncertainties, but in the presence of sensor data overload, heuristic information, limits on real-time applicability and a very wide interval of system uncertainties, the application of adaptive control cannot ensure a high-quality performance.

On the other hand, becoming capable of versatile and diverse reaction, which is typical for human, autonomous and dexterous robots, should be kinematically redundant. This redundancy makes the inverse kinematics problem ill-posed, hence it is more diffucult to find a solution because of the huge amount of computations. Thus, the control problem of the redundant robot should be stated as with humans: use training and learning instead of pure computation.

Therefore, to achieve best performance of the robotic system, a solution of the robot control problem would probably require a combination of conventional approaches with new learning techniques. The application of learning techniques in the field of robot control on the lowest hierarchical level is very important because these techniques can significantly enhance the robotic performance with a priori low level of information about the model of the manipulation robot and the environment. Another important characteristic of learning control in contact tasks is its repetitive nature, which is very important for the process of learning by trial-and-error. Also, the inclusion of the learning concept in robotic control algorithms ensures some specific features as generalizations from multiple examples and the reuse of past experience.

There are many possibilities in the development of more robust robot controllers by utilizing learning systems to identify more accurately the robot kinematics and dynamics, to more efficiently adapt dynamic control parameters to particular tasks, and to more effectively integrate sensory information into the control process. Another possibility for such an application is the field of kinematic calibration of robot arms, utilizing sensing systems to measure positions of the arm end-effector, as the learning system can identify a complex nonlinear model of the robot arm kinematics, which could be used to improve the positioning accuracy of the robot arm itself. The learning systems may also improve the capabilities to execute fine motion operations, such as force control and grasping. Robotic systems offer a promising domain for experimental exploration of learning systems in manufacturing operations, since the practical application of complex robotic systems may require adaptive and learning behaviour in order to achieve the desired functionality.

The problem of control of manipulation robots is considered through a single execution of some technological operation. However, in most cases in industry, robotic operations are inherently of a repetitive nature, where the influence of the working environment is time-varying while the parameters of the robotic system vary in unpredictable ways. In this case, poor performance of the whole system can arise. This is a consequence of the fact that the robot controllers did not use experience acquired through previous repetitive trials, i.e., the robot controllers were based on "short-term memory". The compensation of unmodelled dynamics was not achieved because of the lack of improvement mechanisms of control laws based on repetitive control trials. Hence, important characteristics of new learning control algorithms must be the use of skill refinement, i.e., the ability to gradually improve performance on the same or similar robotic tasks by repeated practice over time. In this way it is possible to achieve autonomous acquisition of the model of the robotic system and active reduction of the system's uncertainties based on "long-term memory". Another important characteristic of new learning laws should be the ability to execute old or new robotic tasks with higher quality (with greater speed, skill and accuracy) based on the generalization of the acquired knowledge.

The robot controllers with the mentioned two properties are called *repetitive controllers* or *training controllers* and they involve two essential phases of work. The first phase is the *training phase* based on repetitive execution of the technological operation with a constant process of knowledge acquisition. In the second phase, called *phase of knowledge association*, the learning process is stopped and, using efficient association from memory, efficient control execution is performed. In this case, the learning control problem is considered as a pattern recognition problem with the input-output mapping based on associative memory data.

For robot learning control on the executive hierarchical level, we can identify four main paradigms:

1 **Iterative - analytical methods**. These methods are based on successive attempts at following the same trajectory with betterment properties. Typically, control input values for each time instant in the trajectory are adjusted

iteratively on the basis of the observed trajectory errors at similar times during the previous attempts. These iterative learning algorithms may be very precise and exhibit rapid convergence. On the other hand, a drawback of such control techniques is that they are applicable only to operations that are repetitive. These algorithms have no capability of generalization on quite different movements.

2 **Tabular methods**. These methods are related to the use of the local, memory-based, nonparametric function approximators (associative content - addressable memories, statistical locally weighted regression or kd-trees). In the case of addressable memory, the robot model is tought by storing experience about command signals and current state coordinates in the memory. Each time a particular set of robot positions, velocities and accelerations is requested, the entire memory has to be searched for the closest experience. In this approach, the problems are extended search time due to the great amount of stored experience, ways to measure similarity, and methods of efficient generalization.

3 **Reactive learning methods**. These are numeric, inductive and continuous methods based on the psychological concept that is necessary to apply a reward immediately after the occurrence of a response in order to increase its probability of reoccurring, while providing punishment if the response decreases the probability. These methods include a component *critic* that is capable of evaluating the response and sendong the necessary reinforcement signal to the robotic control. The problems with this type of learning are the credit assignment and finding optimal decision policy. A robotic agent may have many possible actions it can take in response to a stimulus, and the policy determines which of the available actions the robot should undertake. Two types of reinforcement learning algorithms predominate: the first, *adaptive heuristic critic (AHC) learning* (the process of learning the decision policy for action is separated from learning the utility function the critic uses for state evaluation) and second, *Q-learning* (a single utility Q-function is learned to evaluate both actions and states).

4 **Connectionist methods**. These represent the neural network approach based on distributed processing, where learning is the result of alterations in synaptic weights. Neural networks show great potential for learning the robot structure model and the model of the robotic system together with a great ability for knowledge association and knowledge generalization. They have the capability of being a general approximation tool for complex nonlinear systems. The connectionist approach does not require explicit programming because of general input-output mapping based on fast parallel architecture and sophisticated learning rules.

In the context of robot control on the lowest hierarchical level, the primary aim of neural networks is the implementation of complex input-output kinematic and dynamic relations, i.e., the learning of inverse kinematic and dynamic robot models as parts of robot control algorithms. The important feature of neural networks is the ability to approximate the complete model of the control system, which is important for the compensation of a wide range of the system's uncertainties. By using the properties of association and generalization, neural networks utilize the knowledge acquired on one set of test trajectories to efficiently control a quite different set of working trajectories. Another important role of neural networks in robotics is connected to the robot sensors, the process of sensor fusion and robot perception in order to present important information from the robot, needed for control decisions and algorithms.

The issue of using neural networks in robotics represents a field of intensive research aimed at overcoming some difficulties connected with practical applications. These difficulties are related to the synthesis of practical topologies for complex mappings, forming of fast learning rules, optimal definition of network structures, accuracy of approximation, choice of "good" supervisor for learning, etc.

In the chapters to follow, special attention will be paid to the connectionist method in robotics as the main and basic research paradigm for robotic intelligent control.

Chapter 2

NEURAL NETWORK APPROACH IN ROBOTICS

1. Introduction

Connectionism represents the study of massively parallel networks of simple neuron-like computing units [285],[115]. The computational capabilities of the systems with neural networks are in fact amazing and very promising; they include not only the so-called " intelligent functions" such as logical reasoning, learning, pattern recognition, formation of associations or abstraction from examples, but also the ability to acquire the most skillful performance in the control of complex dynamic systems. They also evaluate a large number of sensors with different modalities, providing noisy and sometimes inconsistent information. Among the useful attributes of neural networks are:

- *Learning.* During the training process, input patterns and the corresponding desired responses are presented to the network, and an adaptation algorithm is used to automatically adjust the network so that it responds correctly to as many patterns as possible in a training set.

- *Generalization.* The generalization takes place if the trained network responds correctly with high probability to input patterns that were not included in the training set.

- *Massive parallelism.* Neural networks have the capability of performing massive parallel processing.

- *Fault tolerance.* In principle, damage to a few links need not significantly impair the overall performance. Network behaviour gradually decays as the number of errors in cell weights or activations increases.

- *Suitability for system integration.* Networks provide a uniform representation of inputs from diverse resources.

- *Suitability for realization in hardware.* Realization of neural networks using VLSI circuit technology is attractive because the identical structure of neurons makes fabrication of neural networks cost-effective. However, the massive interconnection may result in some technical difficulties, such as power consumption and circuitry layout design.

Neural networks consist of many interconnected simple nonlinear systems, which are typically modelled by appropriate activation function. These simple nonlinear elements, called nodes or neurons, are interconnected, and the strengths of the interconnections are denoted by parameters called weights. A basic building block of nearly all artificial neural networks, *and most other adaptive systems,* is the adaptive linear combiner, cascaded by nonlinearity, which provides saturation for decision making. Sometimes, a fixed preprocessing network is applied to the linear combiner to yield nonlinear decision boundaries. In multi-element networks, adaptive elements are combined to yield different network topologies. At input, the adaptive linear combiner receives an analogue or digital input vector $x = [x_0, x_1, \ldots, x_n]^T$ ("input signal", "input pattern") and, using a set of coefficients, the weight vector $w = [w_0, w_1, \ldots, w_n]^T$, it produces the sum s of weighted inputs at its output, together with the bias member b:

$$s = x^T w + b \qquad (2.1)$$

The weighted inputs to a neuron are accumulated and then passed to an activation function, which determines the neuron output:

$$o = f(s) \qquad (2.2)$$

The activation function of a single unit is commonly a simple nondecreasing function like threshold, identity, sigmoid or some other complex mathematical function. A neural network is a collection of interconnected neurons. Neural networks may be distinguished according to the type of interconnection between input and output of the network. Basically, there are two types of networks: feedforward and recurrent. In the feedforward neural network, there are no loops and the signals propagate only in one direction from an input stage through intermediate neurons to the output stage. With the use of a continuous nonlinear activation function, this network represents a static nonlinear map which can be efficiently used as a parallel computational model of continuous mapping. If the network possesses some cycle or loop, i.e., when signals may propagate from the output of any neuron to the input of any neuron, it is then a feedback or recurrent neural network. In a recurrent network the system has an internal state, and thereby the output will also depend on the internal state of the system. Hence, the consideration of recurrent neural networks is connected with the analysis of dynamic systems.

Neural networks are capable storing experimental knowledge through learning from examples. They can be further characterized by their network topology, i.e., by the number of interconnections, the node characteristics that are classified by the type of nonlinear elements used (activation rule) and the kind of learning rules implemented.

The application of neural networks in technical problems consists of two phases:

1) "phase of learning/adaptation/design" is the special phase of learning, modifying and designing of the internal structure of the network, when the network acquires knowledge about the real system as a result of interaction with the system and real environment using the trial and error method, as well as the result of appropriate meta-rules which are inherent to a global network context.

2) "pattern associator phase or associative memory mode" is a special phase when, using the stored associations, the network converges toward the stable attractor or a desired solution.

This chapter gives a basic overview of neural networks and neurocomputing in robotics. We consider the application of network topologies and then focus our discussion on a number of learning principles and rules originating from the supervised and unsupervised modes of learning. In the sequel, we analyze key algorithms of robotic learning, followed by an example of the efficient application of neural networks for robot control on the executive control level.

2. Connectionist Models with Application in Robotics

In the course of research on neural networks, more than 20 neural network models have been developed. Since our attention is focused on the application of neural networks in robotics, we will briefly present some important types of network models that are commonly used in robotic applications. There are multilayer perceptrons (MP), radial basis function (RBF) networks, recurrent version of multilayer perceptron (RMP), Hopfield networks (HN), CMAC networks, ART networks and Kohonen Self-organizing Feature-Mapping Networks (SOFM).

In the study and application of feedforward networks, besides single-layer neural networks, it is convenient to use some more structured networks known as multilayer networks or *multilayer perceptrons*. These networks, with an appropriate number of hidden levels, have received considerable attention because of their better representation capabilities and the possibility to learn highly nonlinear mappings. A typical network topology, which represents a multilayer perceptron (Figure 2.1), consists of an input layer, a sufficient number of hidden layers and the output layer. The following recursive relations define the

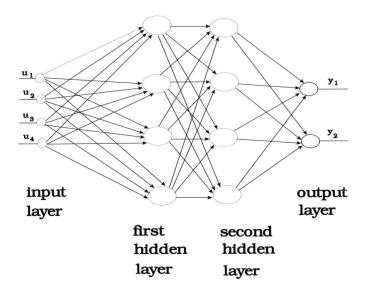

Figure 2.1.   Multilayer perceptron

network with $k + 1$ layers:

$$y_0 = u \tag{2.3}$$
$$y_l = f_l(W_l \bar{y}_{l-1}), \quad l = 1, ..., k \tag{2.4}$$

where y_l is the vector of neuron inputs in the l-layer ($y_k = y$ - output of the $k+1$ - network layer), u is the network input, f_l is the activation function for the l layer, W_l is the weighting matrix between the layers $l - 1$ and l, $\bar{y}_j = [y_j, 1]$ is the adjoint vector y_j. In the previous equation, the bias vector is absorbed by the weighting matrix.

Each layer has an appropriate number of neural units, whereby each neural unit has some specific activation function (usually logistic function sigmoid). The weights of the network are incrementally adjusted according to appropriate learning rules, depending on the task, to improve the system performance. They can be assigned new values in two ways: either determined via some prescribed off-line algorithm - remaining fixed during the operation, or adjusted via a learning process. Although MPs have been widely used for robot learning, they have several drawbacks: tedious tuning of weights and structure and the huge number of learning examples.

An RBF network approximates an input-output mapping by employing a linear combination of radially symmetric functions (see Figure 2.2). The output

y_k is given by:

$$y(u) = \sum_{i=1}^{n} w_i \phi_i(u) \tag{2.5}$$

where:

$$\phi_i(u) = \phi(||u - c_i||) = \phi(r_i) = exp(\frac{-r_i^2}{2\sigma_i^2}), r_i \geq 0, \sigma_i \geq 0 \tag{2.6}$$

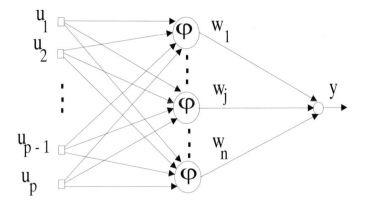

Figure 2.2.   Radial basis function network

This class of neural networks exhibits a diversity of processing units. The input layer plays an important role as an interface between the neural network and its environment. It is built of locally tuned receptive fields, called radial basis functions (RBFs). The RBF repesents the nonmonotonic activation function $\phi(.)$. Theoretical studies have shown that the choice of activation function $\phi(.)$ is crucial for the effectiveness of the network. In most cases the Gaussian RBF given by (2.6) is used, where c_i and σ_i are the selected centres and widths, respectively. In general, an RBF neural network is a feedforward architecture composed of two layers; the first one being built out of the receptive fields while the output layer is a collection of linear processing units. Thus, the main nonlinear effect of neurocomputing resides within the nonlinear RBFs.

One of the earliest sensory connectionist methods that was capable to serve as alternative to the well-known back propagation algorithm was CMAC (Cerebellar Model Arithmetic Computer) [4] (Figure 2.3). The CMAC is a table look-up method for reproducing functions with multiple input and output variables over particular regions of state space. The CMAC topology consists of a three-layer network, one layer is the sensory or command input, the second is the association layer and the third is the output layer. The association layer represents conceptual memory with high dimensionality. On the other hand, the output layer represents the actual memory with low dimensionality. The connections between these two layers are chosen in a random way. The adjustable weights exist only between the association layer and the output layer. Using supervised learning, the training set of patterns is presented and, accordingly, the weights are adjusted. CMAC uses the Widrow-Hoff LMS algorithm [381] as a learning rule. CMAC represents an associative neural network that uses the feature that only a small part of the network influences any instantaneous output. The associative property built into CMAC enables local generalization: similar inputs produce similar outputs while distant inputs produce nearly independent outputs. As a result, we have fast convergence properties. It is very important that today there exists a practical hardware realization using logical cell arrays. The benefits of CMAC modelling are: fast and simple operation, efficient interpolation leading to a global optimum and superposition over input space, and continuous output. Serious drawbacks are massive memory requirement for practical application and its off-line nature, and a very coarse knowledge representation. In addition, the CMAC output is very sensitive to sensor noise.

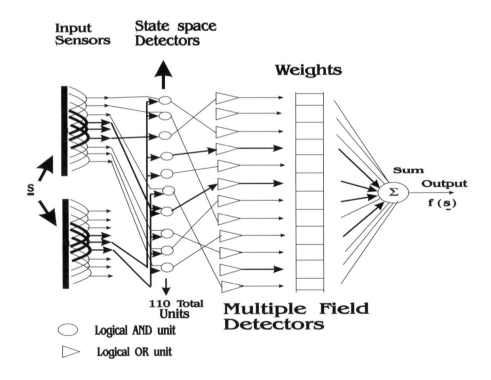

Figure 2.3. Structure of CMAC network

If the network possesses a cycle or loop, then it is a feedback or recurrent neural network. In a recurrent network the system has an internal state, and thereby the output will also depend on the internal state of the system. In the simple case, the state history of each unit or neuron is determined by a differential equation of the form:

$$\dot{y}_i = -a_i y_i + b_i \sum_j w_{ij} u_j \qquad (2.7)$$

where y_i is the state of the $i - th$ neuron, u_j is $j - th$ input to the neuron, w_{ij} are the synaptic weight connections; a_i, b_i are the constants.

These networks represent essentially non-linear dynamic systems with stability problems. There are many different versions of inner and outer recurrent neural networks (recurrent version of multilayer perceptrons) for which efficient learning and stabilization algorithms should be synthesized.

One of the commonly used recurrent networks is the neural network of the Hopfield type [124] (Figure 2.4), which is very convenient for optimization problems. Hopfield introduced a network that employed a continuous nonlinear function to describe the output behaviour of the neurons. The neurons represent an approximation of biological neurons in which a simplified set of important computational properties is retained. This neural network model, which consists of nonlinear graded-response model neurons organized into networks with effectively symmetric synaptic connections, can easily be implemented with electronic devices. The dynamics of this network is defined by the following equation:

$$\dot{y}_i = -\alpha y_i + \beta f_i(\sum_j w_{ij} y_j) + I_i \qquad i = 1, .., n. \qquad (2.8)$$

where α, β are positive constants and I_i is the array of desired network inputs.

A Hopfield network can be characterized by its energy function:

$$E = -\frac{1}{2} \sum_{i=1}^{n} \sum_{j=1}^{n} w_{ij} y_i y_j - \sum_{i=1}^{n} I_i y_i \qquad (2.9)$$

The network will seek to minimize the energy function as it evolves to an equilibrium state. Therefore, one may design a neural network for function minimization by associating variables in an optimization problem with the variables in the energy function. *ART networks* (Figure 2.5) are neural networks based on Adaptive Resonance Theory [44]. An ART network selects its first input as the exemplar for the first cluster. The next input is compared to the first cluster exemplar. It is clustered with the first one if the distance to the first cluster is less than a threshold. Otherwise, it is the exemplar for a new cluster. This procedure is repeated for all the subsequent inputs. The ART networks provide a

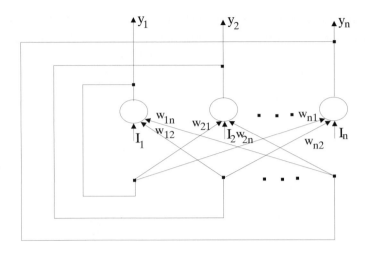

Figure 2.4.   Hopfield network

fertile ground for gaining a new understanding of biological intelligence. They also suggest novel computational theories and real-time adaptive architectures with promising properties for tackling some of the outstanding problems in computer science.

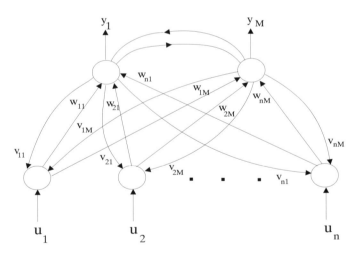

Figure 2.5.   ART network

The principal goal of the Kohonen self-organizing feature-mapping networks (SOFM) [190] is to transform an incoming signal pattern of arbitrary dimension into a one or two-dimensional discrete map and perform this transformation adaptively in a topologically ordered fashion. A lot of activation patterns are presented to the network, one at a time. Each presentation activates a corresponding localized group of neurons in the output layer of the network.

The essential gradients of the neural networks embodied in such an algorithm include a one or two dimensional lattice of neurons (Figure 2.6) that compute simple discriminant functions of inputs received from the input of an arbitrary dimension. For application, most essential is the mechanism that compares these discriminant functions and selects the neurons with the highest discriminant function value. The weight adaptation is achieved by an adaptation process that enables the activated neurons to increase their discriminant functions values in relation to the input signals. Depending on the application of interest, the response of the network can be either the index of the winning neuron (position in the lattice) or the synaptic weight vector that is closest to the input vector in a Euclidean space. It is important to notice that the SOFM network displays the important statistical characteristics of the input. In this way SOFM represents the set of synaptic weight vectors in the output space that provide a good approximation to the input space.

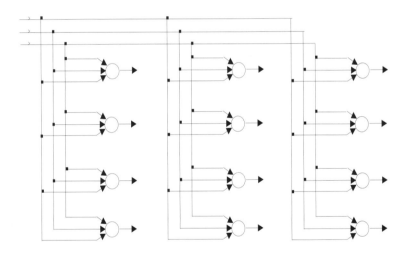

Figure 2.6.   SOFM network

All the proposed neural networks can be classified according to their generalization capability. CMAC is a local generalizing NN, while MLP and recurrent MLP are suitable for global generalization. RBF networks can be placed in between. The choice of network depends on the requirement for local generalization. When a strong local generalization is needed, CMAC is most suitable. For global generalization, MLP and recurrent MLP provide good alternatives when combined with an improved weight adjustment algorithm.

3. Learning Principles and Learning Rules Applied in Robotics

Adaptation (or machine learning) deals with finding weights (and sometimes network topology) that will produce a desired behaviour. Neural networks are inherently plastic structures that call for substantial learning. In fact, the learning occurs under different learning conditions and can be completed under various types of interaction with the environment. Generally, there are three main types of learning, namely

1 Supervised learning

2 Reinforcement learning

3 Unsupervised learning

These three modes of learning are listed in order of the increasing level of challenge they pose to the neural network.

Usually, the learning algorithm works from training examples, whereby each example incorporates correct input-output pairs *(supervised learning).* . This learning form is based on the acquisition of mapping by the presentation of training exemplars (input-output data). The neural network has to reproduce (approximate) these pairs to the highest possible extent.

In contrast to supervised learning, *reinforcement learning* considers the improvement of system performance by the evaluation of some realized control action, which is included in the learning rules. Reinforcement learning shows an interesting behaviour and assumes a number of different scenarios. Overall, the reinforcement means that the network is provided with a global signal about the network's performance. This signal acts as a penalty (or reward) mechanism. The higher its value, the more erroneous the behaviour of the network. The reinforcement can also be a function of a global target. The available reinforcement can be spatial or temporal, or both. Spatial reinforcement occurs when we are provided not with the individual target values but with a form of its scalar aggregate r. Temporal reinforcement learning occurs when the network is updated on its performance after a certain period of time. A relevant example is when the time series produced by the network is evaluated over a certain time slice, T, rather than being monitored continuously.

Finally, the mode of unsupervised learning makes no provisions for any type of supervision and the network has to reveal and capture essential dependences occurring in the data set autonomously. The *unsupervised learning* process in connectionist learning represents learning when the processing units respond only to interesting patterns at their inputs, based on its internal learning function. The topology of the network during the training process can be fixed or variable based on evolution and regeneration principles. Quite commonly, unsupervised learning is referred to as clustering, and covers a significant number of pertinent algorithms, including hierarchical clustering, C-Means, Fuzzy C - Means, etc.

Various iterative adaptation algorithms that have so far been proposed, being essentially designed in accord with the *minimal disturbance principle:* are adapted to reduce the output error for the current training pattern, with minimal disturbance to the responses already learned. Two principal classes of algorithms can be distinguished:

Error-correction rules, which alter the weights of a network to correct the error in the output response to the present input pattern;

Gradient-based rules, which alter the weights of a network during each pattern presentation by gradient descent with the objective of reducing mean-square errors, averaged over training patterns.

The error-correction rules for networks often tend to be ad-hoc in nature. They are most often used when training objectives are not easily quantified, or when a problem does not lend itself to tractable analysis (for instance, networks that contain discontinuous functions, e.g., signum networks).

Gradient adaptation techniques are intended for minimization of the mean-square error associated with an entire network of adaptive elements:

$$e^2 = \sum_{t=1}^{T} \sum_{i=1}^{N_y} [e_i(t)]^2 \tag{2.10}$$

where $e_i^2(t)$ is the square error for particular patterns.

The most practical and efficient algorithms typically work with one pattern presentation at a time. This approach is referred to as *pattern learning,* in contrast to *batch learning,* in which weights are adapted after presentation of all the training patterns (true *real-time learning* is similar to pattern learning, but it is performed with only one pass through the data). Similar to the single-element case, in place of the true MSE function, the instantaneous sum squared error $e^2(t)$ is considered, which is the sum of the square errors at each of the N_y outputs of the network:

$$e^2(t) = \sum_{i=1}^{N_y} [e_i(t)]^2 \tag{2.11}$$

The corresponding instantaneous gradient is:

$$E = \widehat{\nabla}(t) = \frac{\partial e^2(t)}{\partial w(t)} \tag{2.12}$$

where $w(t)$ denotes a vector of all weights in the network. The steepest descent with the instantaneous gradient is a process presented by

$$
\begin{aligned}
w(t+1) &= w(t) + \Delta w(t) \\
\Delta w(t) &= \mu(-\widehat{\nabla}(t))
\end{aligned} \tag{2.13}
$$

The most popular method for estimating the gradient $\widehat{\nabla}(t)$ is the *back propagation algorithm(BP)*.

There are several powerful learning algorithms for feedforward networks, but the most commonly used is the *back propagation algorithm* [285]. The back propagation algorithm, as a typical supervised learning procedure, adjusts weights in the local direction of greatest error reduction (steepest descent gradient algorithm) using the square criterion between the real network output and desired network output.

Basic analysis of the algorithm application will be shown using a three-layer perceptron (one hidden layer with sigmoid functions in the hidden and output layers). The main relations in the training process for one input-output pair $p = p(t)$ are given by the following relations:

$$
\begin{aligned}
s_2^p &= W_{12}^{p}{}^T u_1^p & s_2^p \epsilon R^{L_1} & \tag{2.14} \\
o_{2a}^p &= 1/(1 + exp(-s_{2a}^p)) & a = 1, ..., L_1 \quad o_{20}^p = 1 & \tag{2.15} \\
s_3^p &= W_{23}^{p}{}^T o_2^p & s_3^p \epsilon R^{N_y} & \tag{2.16} \\
o_{3b}^p &= 1/(1 + exp(-s_{3b}^p)) & b = 1, ..., N_y & \tag{2.17} \\
y_c^p &= o_{3c}^p & c = 1, .., N_y & \tag{2.18}
\end{aligned}
$$

where s_2^p, s_3^p are the input vectors of the hidden and output layers of the network; o_2^p, o_3^p are the output vectors of the hidden and output layers; $W_{12}^p = [w_{12ij_{N_u+1 \times L1}}(t)]$, $W_{23}^p = [w_{23ij_{L1+1 \times N_y}}(t)]$ are the weighting factors; w_{tuij} is the weighting factor which connects the neuron j in the layer t with the neuron i in the output layer u; u_1^p represents the input vector ($u_{10}^p = 1$; N_u - number of inputs); y^p is the output vector (N_y - number of outputs).

The square error criterion can be defined as:

$$
E = \sum_{p \in P} E^p = 0.5 \sum_{p \in P} | \hat{y}^p - y^p |^2 \tag{2.19}
$$

where $\hat{y}^p$ is the desired value of the network output; y^p is the output value of the network; E^p is the value of the square criterion for one pair of input-output data; P is the set of input-output pairs.

The corresponding gradient component for the output layer is:

$$
\frac{\partial E}{\partial w_{23ij}} = \sum_{p \in P} \frac{\partial E^p}{\partial w_{23ij}} = \sum_{p \in P} \frac{\partial E_p}{s_{3i}^p} \frac{\partial s_{3i}^p}{\partial w_{23ij}} = - \sum_{p \in P} \delta_{3i}^p o_{2j}^p \tag{2.20}
$$

$$
\delta_{3i}^p = (\hat{y}_i^p - y_i^p) df_{3i}^p / ds_{3i}^p = (\hat{y}_i^p - y_i^p) f_{3i}'(s_{3i}^p) \tag{2.21}
$$

where f_{gi} is the activation function for the neuron i in the layer g.

For the hidden layer, gradient component is defined by:

$$
\begin{aligned}
\frac{\partial E}{\partial w_{12ij}} &= \sum_{p \in P} \frac{\partial E^p}{\partial w_{12ij}} = \sum_{p \in P} \frac{\partial E_p}{s_{2i}^p} \frac{\partial s_{2i}^p}{\partial w_{12ij}} \\
&= \sum_{p \in P} \sum_r \frac{\partial E^p}{\partial s_{3r}^p} \frac{\partial s_{3r}^p}{\partial o_{2i}^p} \frac{\partial o_{2j}^p}{\partial s_{2i}^p} \frac{\partial s_{2i}^p}{\partial w_{12ij}} \\
&= -\sum_{p \in P} \sum_r \delta_{3r}^p w_{23ri} f_{2i}'(s_{2i}^p) u_{1j}^p \\
&= -\sum_{p \in P} \delta_{2i}^p u_{1j}^p \qquad\qquad (2.22) \\
\delta_{2i}^p &= \sum_r \delta_{3r}^p w_{23ri} f_{2i}'(s_{2i}^p) \qquad\qquad (2.23)
\end{aligned}
$$

Based on previous equations and starting from the output layer and going back, the error back propagation algorithm is synthesized. The final version of the algorithm, with modification of the weighting factors, is defined by the following relations:

$$
\delta_{3i}(t) = (\hat{y}_i(t) - y_i(t)) f_{3i}'(s_{3i}(t)) \qquad\qquad (2.24)
$$

$$
\Delta w_{23ij}(t) = -\eta \frac{\partial E}{\partial w_{23ij}} = \eta \delta_{3i}(t) o_{2j}(t) \qquad\qquad (2.25)
$$

$$
\delta_{2i}(t) = \sum_r \delta_{3r}(t) w_{23ri}(t) f_{2i}'(s_{2i}(t)) \qquad\qquad (2.26)
$$

$$
\Delta w_{12ij}(t) = -\eta \frac{\partial E}{\partial w_{12ij}} = \eta \delta_{2i}(t) u_{1j}(t) \qquad\qquad (2.27)
$$

$$
w_{23ij}(t+1) = w_{23ij}(t) + \Delta w_{23ij}(t) \qquad\qquad (2.28)
$$

$$
w_{12ij}(t+1) = w_{12ij}(t) + \Delta w_{12ij}(t) \qquad\qquad (2.29)
$$

where η is the learning rate.

Numerous modifications are also used to speed up the learning process in back propagation algorithm. An important extension is the *momentum technique*, which involves a term proportional to the weight change from the preceding iteration:

$$
\begin{aligned}
w(t+1) &= w(t) + \Delta w(t) \\
\Delta w(t) &= (1 - \eta) \cdot \mu(-\hat{\nabla}(t)) + \eta \cdot \Delta w(t-1) \qquad\qquad (2.30)
\end{aligned}
$$

The momentum technique serves as a low-pass filter for gradient noise and it is useful in situations when a clean gradient estimate is required, for example when a relatively flat local region in the mean square error surface is encountered.

All gradient-based methods are subject to convergence on local optima. The most common remedy for this is the sporadic addition of noise to the weights or gradients, as in *simulated annealing* methods. Another technique is to re-train the network several times using different random initial weights until a satisfactory solution is found. The back propagation adapts the weights to seek the extremum of the objective function whose domain of attraction contains the initial weights. Therefore, both the choice of initial weights and the form of the objective function are critical to the network performance. The initial weights are normally set to small random values. Experimental evidence suggests that the initial weights in each hidden layer are to be chosen in a quasi-random manner, which ensures that at each position in a layer's input space the outputs of all but a few of its elements are saturated, while ensuring that each element in the layer is unsaturated in some region of its input space.

There are a number of different learning rules for speeding up the convergence process of the back propagation algorithm. One interesting method is the use of recursive least square algorithms and Extended Kalman approach, instead of gradient techniques [167].

The training procedure for the RBF networks involves several important steps:

Step 1: Group the training patterns in M subsets using some clustering algorithm (k - means clustering algorithm) and select their centres c_i.

Step 2: Compute the widths, $\sigma_i, (i = 1, .., m)$, using an heuristic method (p - nearest neighbour algorithm) .

Step 3: Compute the RBF activation functions $\phi_i(u)$ for the training inputs.

Step 4: Compute the weight vectors by least squares.

Using supervised learning, the training set of patterns is presented for CMAC networks and, accordingly, the weights are adjusted. CMAC uses the Widrow-Hoff LMS algorithm [381] as the learning rule.

ART networks belong to the class of unsupervised learning networks. They are stable because new input patterns do not wash away previously learned information. They are also adaptive because new information can be incorporated until the full capacity of the architecture is utilized.

If an input of an ART network is clustered with the $j-th$ cluster, the weights of the network are updated according to the following formulae

$$w_{ij}(t+1) \quad = \quad \frac{v_{ij}(t)u_i}{0.5 + \Sigma_{i=1}^n v_{ij}(t)u_i} \tag{2.31}$$

$$v_{ij}(t+1) \quad = \quad u_i v_{ij}(t) \tag{2.32}$$

where $i = 1, 2, ..., M$.

The main steps of the adaptation process for SOFM networks are summarized as follows [190]:

1 *Initialization*. Choose random values for the initial weight vectors $w_j(0)$.

2 *Sampling*. Draw a sample x from the input distribution with a certain probability, the vector x represents the sensory signal.

3 *Similarity Matching*. Find the best-matching (winning) neuron $i(x)$ at the time n, using the minimum-distance Euclidean criterion:

$$i(x) = argmin\|x(n) - w_j\|, j = 1, 2, ..., N \qquad (2.33)$$

4 *Updating*. Adjust the synaptic weight vectors of all neurons, using the update formula

$$w_j(n+1) = \begin{cases} w_j(n) + \eta(n)[x(n) - w(j)(n)], & j \in \Lambda(n) \\ w_j(n) & \text{otherwise} \end{cases} \qquad (2.34)$$

where $\eta(n)$ is the learning-rate parameter and $\Lambda(n)$ is the neighbourhood function entered around the winning neuron $i(n)$; both $\eta(n)$ and $\Lambda(n)$ are varied dynamically during learning for best results.

5 *Continuation*. Continue with step 2 until no noticeable changes in the feature map are observed.

4. Neural Network Issues in Robotics

One of the main goals in the research on modern intelligent robotic systems represents the possibility that robots learn from their experience. As we know, a robot learns from what is happening rather than from what is expected. Hence, in the case where self-adaptive and autonomous capabilities are required, the application of neural networks in robot control has a great chance of success. Neural networks are able to recognize the changes in the robot environment conditions and then react to them, or, make decisions based on changing manufacturing events.

The area of possible applications of neural networks in robotics [28] [126], [195],[166],[135],[347], [211],[91],[147] includes various purposes such as vision systems, appendage controllers for manufacturing, tactile sensing, tactile feedback gripper control, motion control systems, situation analysis, navigation of mobile robots, solution of the inverse kinematic problem, sensory-motor coordination, generation of limb trajectories, learning visuomotor coordination of the robot's arm in 3D, control of biped gait, etc. All these robotic tasks can be categorized according to the type of hierarchical control level of the robotic system, i.e., neural networks can be applied at the strategic control level (task planning), at the tactical control level (path planning) and at the executive control level (path control). All these control problems on different hierarchical levels can be formulated in terms of optimization or pattern association problems. For example, autonomous robot path planning and stereo vision for task planning can be formulated as optimization problems, while, on the other hand, sensor/motor control, voluntary movement contrl and cerebellar model articulation controller can be formulated as pattern association tasks. Precisely, for pattern association tasks, neural networks in robotics can have the role of function approximation (modelling of input/output kinematic and dynamic relations) or the role of pattern classification, necessary for control purposes.

4.1   Kinematic robot learning by neural networks

It is well known in robotics that control is applied at the level of the robot's joints, while the desired trajectory is specified through the movement of the end-effector. Hence, a control algorithm requires the solution of the inverse kinematic problem for a complex nonlinear system (connection between internal and external coordinates) in real time. However, in general, the path in Cartesian space is often very complex and location of the arm end-effector cannot be efficiently determined before the movement is actually made. Also, the solution of the inverse kinematic problem is not unique because of the fact that in the case of redundant robots there may be an infinite number of solutions. In this case, conventional methods of solution consist of closed-form methods and iterative methods. These are either limited only to a class of simple non-redundant robots or are time-consuming, and the solution may diverge due to a bad initial guess. We shall refer to this method as *position-based inverse kinematic control*. The *velocity-based inverse kinematic control* directly controls the joint velocity (relation between external and internal velocities by Jacobian matrix). The velocity-based inverse kinematic control is also called inverse Jacobian control.

In the area of position-based inverse kinematic control, various methods have been proposed to solve this problem [183], [278], [333]. The basic idea common to all these algorithms is the use of the same topology of neural network (multilayer perceptron) and the same learning rule: back propagation algorithm. Although it has been demonstrated that back propagation algorithms work in the case of robots with a small number of DOFs, they may not perform in the same way for robots with six DOFs. In fact, the problem is that these methods are naive, i.e., in the design of the neural network topology some parts of the knowledge about the kinematic robot model have not been incorporated. One solution is to use a hybrid approach, i.e., a combination of the neural network approach and classic iterative procedures. In the first phase of training, the multilayer neural network is used with the desired position and orientation vector as input and the corresponding joint values as the output of the network (supervised learning). In the second phase (operation phase), the trained neural network is connected with the iterative method. The neural network then gives an approximate solution. which represents the initial guess for the iterative method. The iterative method gives the final solution for joint coordinates within the specified tolerance.

A special algorithm for solving the position-based kinematic problem represents the application of polynomial neural networks [47]. The polynomial neural networks are based on Ivakhnenko's Group Method of Data Handling (GMDH) [145]. GMDH is a procedure for the automatic construction of a polynomial in order to obtain an appropriate input-output relationship by incorporating a series of simple intermediate second-order polynomials. These intermediate models are systematically constructed step - by - step by passing through the network of successive layers to achieve the empirical extraction of a polynomial combination of variables which best describes the controlled system. It requires a small number of learning examples.

Guo and Cherkassky [107] proposed a solution using the Hopfield model of neural network. Their solution is not directly developed for inverse kinematic relations but instead couples the Hopfield net with Jacobian control techniques.

It is well known that the inverse kinematics function of the robot arms is a multi-valued and discontinuous function. Hence, it is diffucult for a well-known multi-layer neural network to approaximate such a function. Some researchers [267] proposed a modular neural network architecture for inverse kinematics learning. The precise inverse kinematics model is learned by the appropriate switching of multiple neural networks. Each expert neural network approximates the continuous region of the inverse kinematic function.

In the velocity-based kinematic approaches, the neural network has the task of mapping the external velocity into joint velocity. In paper [392], a very interesting approach has been proposed based on using context-sensitive networks. It is an alternative approach to the reduction of complexity, as it proposes partition of the network input variables into two sets. One set (context input) acts as the input to a context network. The output of the context network is used to set up the weights of the function network. The function network maps the second set of input variables (function input) to the output. The original function to be learned is decomposed into a parameterized family of functions, each of which is simpler than the original one, thus being easier to learn. The standard back-propagation learning algorithm can be easily extended to work in a context-sensitive network.

Recently, some researchers have begun considering the use of neural networks for the kinematic control of humanoid walking [197], [157], exactly for the generation of trajectories (gait) of humanoid robots. This approach makes possible the learning of new gaits, which are not weighted combinations of predefined biped gaits. Kurematsu [197] proposed a multi-layered network by using the centre of gravity concept in trajectory generation. For example, Juang and Lin [157] used the algorithm of back propagation through time to synthesize the biped robot gait. Due to the high number of DOFs, it is difficult to obtain a high nonlinear model of the biped. Hence, the complex inverse dynamic computations were eliminated by using a linearized inverse biped model. The neural controller is a three-layer feedforward network. The simulation results show that the neural network, as open-loop controller. can generate control sequences to drive the biped along a prespecified trajectory. This algorithm can also be used for slope surface training.

In recent years, many neural networks (perceptron, RBF networks or recurrent networks) have been applied to the velocity-based kinematic control of redundant robot manipulators [367], [208], [233], [334], [335], [64], [383], [29], [241] , [407]. A new recurrent neural network, called the dual network, is presented in a paper [383]. Unlike feedforward neural networks, most recurrent neural networks do not use off-line supervised learning and are thus more suitable for real-time control in uncertain environments. Mao and Xsia [233] investigated the neural network approach to solve the inverse kinematics problem of redundant manipulators in an environment with obstacles. Their research work has proved the effectiveness of the neural network for trajectory planning and control of robots. Ding and Wang [64] and Wang et al. [367] developed two recurrent neural networks for kinematic control of redundant manipulators that can deal with the case when the Jacobian matrix is in or near singular configuration. Xia and Wang [383] proposed the one-layer dual network that provides a new parallel distributed model for computing the inverse kinematic model. Compared with supervised learning neural networks for robot kinematic control, this approach eliminates the need for training and guarantees rapid convergence. On the other hand, compared with other recurrent neural networks, this approach explicitly minimizes the norm of joint velocity and therefore performs better in terms of joint velocity norm, as well as of position and velocity errors.

In [241] a Radial Basis Function Network (RBFN) approach for fast inverse kinematics computation and effective singularities prevention of redundant manipulators is presented. It is based on computation of the inverse kinematics using a novel RBFN approach, and also on a nonconventional implementation of a novel geometric framework for the singularities prevention of redundant manipulators. Such a framework permits the development of fast algorithms, implementable in real time. Here, some characterizing matrices, representing a number of geometrical concepts, are first established in order to obtain a meaningful performance index and a null space vector for singularities avoidance/prevention and safe path generation. Next, this null space vector is properly included in the off-line training of the RBFN utilized for the computation of the inverse kinematics. Once the RBFN is properly trained, it is readily available for on-line use in truly real-time and fast computation of the inverse kinematics and safe path generation. Here, the effectiveness of the proposed RBFN approach is demonstrated by the successful singularities prevention of a planar redundant manipulator.

Generally, the main problem in all kinematic approaches is the accuracy of tracking a predetermined robot trajectory. As is known, in most kinematic connectionist approaches, the kinematic input/output mapping is learned off-line and then control is attempted. However, it is necessary to examine the proposed solutions by learning control of manipulation robots in real time, because robotic syatems are complex dynamic systems.

4.2 Dynamic robot learning by neural networks

As a solution in the context of robot dynamics learning, the neural network approaches provide the implementation tools for complex input/output relations of robot dynamics without analytic modelling. Perhaps the most powerful property of neural networks in robotics is their ability to model the whole controlled system itself. In this way the connectionist controller can compensate for a wide range of robot uncertainties. It is important to notice that the application of connectionist solutions to robot dynamics learning is not limited to noncontact tasks. It is also applicable to the essential contact tasks, where inverse dynamic mapping is more complex, because the dependence of contact forces is included.

One of the main goals of dynamics learning methods is to find a solution for the inverse dynamic problem. Let us explain the inverse dynamic problem of robot control in a computational framework. There are causal relations between the robot driving torques and the resulting robot movement coordinates. Let $P(t)$ denote the time history of driving torque and $q(t)$ denote the time history of the robot internal coordinates during the trajectory. We can also denote the causal relation between P and q using the functional F, i.e., $F(P(.)) = q(.)$. If we want the robot to track a desired trajectory q_d, the problem of generating a desired driving torque P_d which realizes q_d, is equivalent to finding an inverse of the functional F. However, the inverse dynamic mapping in this case is more complex than the kinematic mapping, because the dependence of robot internal coordinates, velocities and accelerations is included.

The proposed neural network models can be regarded as examples of the autonomous driving torque generator. In this case, the connectionist solution is commonly used as a feedforward controller in the control algorithm. Here, the feedback controller serves as a robust controller with the aim of achieving low errors and the performance of high-quality learning, because the feedforward controller alone is not sufficient for accurate tracking.

Several neural network models and learning schemes have recently been applied to design robot controllers. One of the main distinctions between these methods is the extent of the knowledge of dynamic models that is used in the design procedures.

Some methods use the complete available information on robot model in the design procedure. Out of this group of methods we shall mention as fundamental

Kawato's approach [173], which uses a neural structure based on a nonrecurrent single-layer feedforward neural network. The inputs to this neural unit represent some specific nonlinear functions which are present in the developed form of the inverse dynamic model of manipulation robot. In the training phase, a conventional feedback PID-controller provides a moderately skillful function of the system. The feedback signals serve as the error signals for the single-layer neural identifier.

Kawato's approach is a deterministic technique, because it uses any piece of a priori knowledge about the dynamic structure of the controlled object and thus has better scale-up properties. This approach, however, has several drawbacks related primarily to the inherent complexity of the implementation of a complete model of robot dynamics, which demands large computational power in real time. Second by the way of determining an appropriate set of nonlinear functions is a non-trivial and non-unique task. The limitation of this approach is evident in the case where the exact form of the dynamics equation is not known a priori, i.e. the system is not capable of learning how to handle a completely unknown manipulator. Moreover, one of the main drawbacks is the problem of generality, because the complexity of the formulation of nonlinear subsystems is directly dependent on the features of each specific model.

A special group of dynamic connectionist approaches are the methods that use the "black-box" approach in the design of neural network algorithms for robot dynamic control. The "black-box" approach does not use any a priori experience and knowledge about the inverse dynamic robot model. In this case it is a multilayer neural network with a sufficient number of hidden layers. All we need to do is to feed the multilayer neural network the necessary information (desired positions, velocities and accelerations at the network input and desired driving torque at the network output) and allow it to learn by test trajectory. In paper [268], a nonlinear neural compensator which incorporates the idea of the computed torque method, is presented. Although the pure neural network approach without any knowledge of robot dynamics may be promising, it is important to notice that this approach is not practical because of the high dimensionality of input-output spaces. In [26], use is made of the principle of functional decomposition to simplify robot dynamics learning. This method includes a priori knowledge about the robot dynamics which, instead of being particular knowledge corresponding to a certain type of robot model, incorporates a common knowledge of robot dynamics. In this way, the unknown input-output mapping is decomposed into simpler functions, which are easier to learn because of smaller domains. The main achievement of this approach is an exponential reduction in the size of input space, which is important for practical solutions of robot controllers. Also, by this method, the relative error instead of the absolute error is minimized, which provides a better balance between the control torque and derived external acceleration. In [167], similar ideas on the development of a fast learning algorithm were used, with the decomposition on the level of internal robot coordinates, velocities and accelerations.

The application of the connectionist approach in robot dynamics control can be classified according to the type of learning into two main groups: neurocontrol via supervised learning and neurocontrol via unsupervised learning .

For the first class of neurocontrol a teacher is assumed to be availabl and capable of teaching the required control performance. It is learning from examples provided by some knowledgable external teacher. This is a good approach in the case of a human-trained controller, since it can be used to automate a previously human-controlled system. However, in the case of automated linear and nonlinear teachers, the teacher's design requires an a priori knowledge of the dynamics of the robot under control. The structures of supervised neurocontrol involves three main components, namely a teacher, the trainable controller, and the robot under control [102]. The teacher can be either a human controller or an automated controller (algorithm, knowledge-based process, etc.). The trainable controller is a neural network appropriate for supervised learning, prior to training. Robot states, measured by specialized sensors, are sent to both the teacher and the trainable controller. During the control of the robot by the teacher, the control signals and state variables of the robot are sampled and stored for neural controller training. After successful training the neural network has learned the right control action and replaces the teacher in controlling the robot.

There are several different principal architectures for supervised robot learning. In the *specialized learning architecture* (Figure 2.7), the neural network is tuned by the difference between the desired response and actual response of the system. As another solution, *generalized learning architecture* (Figure 2.8) is proposed. The network is first trained off-line, based on control error, until good convergence properties are achieved, and then put to a real-time feedforward controller, where the network continues its adaptation to the system changes according to specialized learning procedures.

Generally, a possible way to avoid the problem of error propagation through the controlled system is a combination of both learning architectures. It is better to use general training first, to learn the approximate behaviour of the controlled system, followed by specialized training to achieve fine tuning of the network in the current operating regime. General learning architecture has a tendency to create better initial weights for specialized learning. In this way, the time duration of the learning process is decreased. Another advantage is that the neural network can adapt more easily if the current operating points of the systems are changed or new ones are added.

However, the most appropriate learning architectures for robot control are *feedback-error learning architecture and adaptive learning architecture*. The feedback-error learning architecture (Figure 2.9) is an exclusively on-line architecture for robot control, which enables a simultaneous process of learning

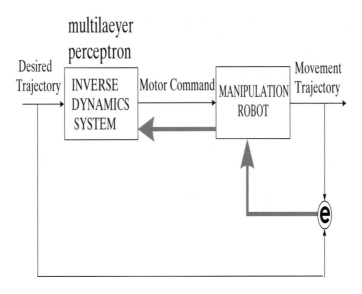

Figure 2.7.   Specialized learning architecture

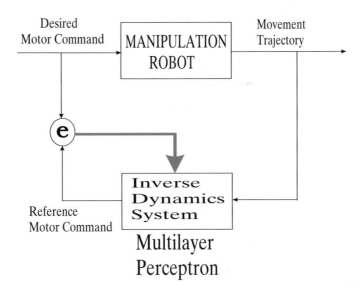

Figure 2.8.   Generalized learning architecture

and control. The primary interest is the learning of an inverse dynamic model of the robot mechanism for tasks with holonomic constraints, where exact robot dynamics is generally unknown. The neural network, as part of the feedforward control, generates the necessary driving torques at robot joints as a nonlinear mapping of the robot's desired internal coordinates, velocities and accelerations:

$$P_i = g(w_{jk}^{ab}, q_d, \dot{q}_d, \ddot{q}_d) \qquad i = 1, .., n. \qquad (2.35)$$

where $P_i \epsilon R^n$ is the joint driving torque generated by the neural network; w_{jk}^{ab} are the adaptive weighting factors between the neuron j in the $a - th$ layer and the neuron k in the $b - th$ layer; g is the nonlinear mapping.

According to the integral model of robotic systems, the decentralized control algorithm with learning has the form

$$u_i = u_{ii}^{ff} + u_{ii}^{fb} \qquad i = 1, .., n. \qquad (2.36)$$

$$u_i = f_i(q_d, \dot{q}_d, \ddot{q}_d, P) - KP_{ii}\varepsilon_i - KD_{ii}\dot{\varepsilon}_i - KI_{ii} \int \varepsilon_i dt \qquad i = 1, .., n. \qquad (2.37)$$

where f_i is the nonlinear mapping, which describes the nature of robot actuator model; $KP \epsilon R^{n \times n}$, $KD \epsilon R^{n \times n}$, $KI \epsilon R^{n \times n}$ are the position, velocity and integral local feedback gains, respectively; $\varepsilon \epsilon R^n$ is the feedback error. In the

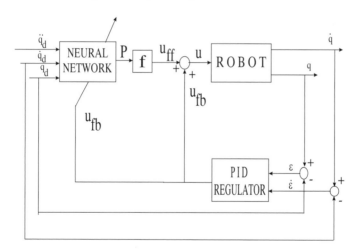

Figure 2.9. Feedback-error learning architecture

process of training we can use the feedback control signal:

$$e_i = u_i^{fb} \qquad i = 1, .., n. \qquad (2.38)$$

where $e_i^{bp} \epsilon R^n$ is the output error for the back propagation algorithm.

In unsupervised neural learning control, no external teacher is available and the dynamics of the robot under control is unknown and/or involves severe uncertainties. There are several different principal architectures for unsupervised robot learning.

One of the most important and interesting approaches is *reinforcement learning* (Figure 2.10). Reinforcement learning differs from supervised learning. Supervised learning alone is not adequate for learning from interaction with the environment. In this case, there is no explicit teacher, but there is a direct sensory-motor connection to the environment. Reinforcement learning is learning what to do, how to map situations to actions so as to maximize a numerical reward signal. In contrast to the supervised learning paradigm, the role of the teacher in reinforcement learning is more evaluative than instructional. The learner is not told which actions to take, but has to discover which actions yield the most reward by trying them. The teacher provides the learning system with an evaluation of the system performance of the robot task according to a certain criterion. The aim of this learning system is to improve its performance by generating appropriate outputs.

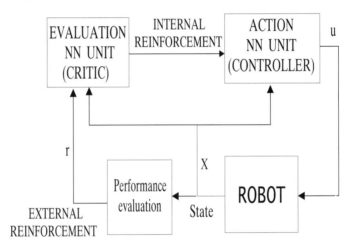

Figure 2.10.   Reinforcement learning architecture

In [106], a stochastic reinforcement learning approach, with application in robotics for learning functions with continuous outputs, is presented. The learning system computes real-valued output as some function of a random activation, generated using normal distribution. The parameters of normal distribution are the mean and standard deviation that depend on current input patterns. The environment evaluates the unit output in the context of input patterns and sends a reinforcement signal to the learning system. The aim of learning is to adjust the mean and standard deviation so as to increase the probability of producing the optimal real value for each input pattern. The unit having the above properties is called a stochastic real-valued (SRV) unit. The author presents an experiment in robot control (robot with 3 DOFs) with a network of units for a positioning task. The task is defined as: Given the current position of the arm in joint-space coordinates, the change in the position of the arm over the last time step, and a desired x-coordinate location for the end point x, determine the new joint-space coordinates for the arm that will result in the end point of the arm having x as its x-coordinate. As we see, this learning paradigm is not intended for continuous tracking of robot trajectories. In other words, if the network is given as input a desired final location and robot configuration close to the final robot configuration for that location input, it should produce the final robot configuration as output. There are seven inputs and three outputs of the network. Also, the network has two types of units: an appropriate number of back-propagation units (in this case 16) and 3 SRV units. The SRV units send correction signals to back-propagation units to update their weights. The environment returns a reinforcement signal, which is a function of the averaged position error. The results show that this approach is capable of successfully computing position controls for the robot manipulators. This trained network also shows interesting properties, such as the existence of many limit points, which correspond to stable solutions of the positioning task. Generally, a drawback of this approach, in comparison with supervised learning is the slow learning rate. In a similar fashion, Song and Sun [313] used a reinforcement learning control for the dynamic control of a two-link manipulator. The learning controller can achieve the desired position and force performances via repetitive learning. Practical experiments on a two-link direct-drive robot demonstrated the capability of this approach to eliminate load disturbance applied to the robot manipulator. However, because of the slow learning process, the need for a more efficient learning structure to expedite the convergence of the learning phase remains for future research.

The important characteristics of connectionist dynamic learning methods in robotics is the fact that a neural network represents a part or upgrade of conventional non-adaptive and nonlinearly parameterized or linearly parameterized adaptive control laws applied in robotics. For example, connectionist structures were used for feedforward dynamic control [168],[323], indirect adaptive control laws [43], [203], sliding mode control laws [84], or computed torque control law [268].

In [271], an approach and a systematic design methodology to adaptive motion control based on neural networks for high-performance robot manipulators is presented. The neurocontroller includes a linear combination of a set of off-line trained neural networks (bank of fixed neural networks), and an update law of the linear combination coefficients to adjust the robot dynamics and payload uncertain parameters. A procedure is presented to select the learning conditions for each neural network in the bank. The proposed scheme, based on fixed neural networks, is computationally more efficient than using the learning capabilities of the neural network to be adapted as that used in feedback architectures that need to propagate back control errors through the model (or network model) to adjust the neurocontroller. A practical stability result for the neurocontrol system is given. That is, it has been proved that the control error converges asymptotically to the neighbourhood of zero, the size is evaluated and depends on the approximation error of the neural network bank and the design parameters of the controller. In addition, a robust adaptive controller for neural network learning errors is proposed, using a sign or saturation switching function in the control law, which leads to global asymptotic stability and zero convergence of control errors. The presented simulation results showed the practical feasibility and performance of the proposed approach in robotics.

An interesting concept of Leahy, Johnson and Rogers [203] employs neural networks to recognize variations in the manipulator payload. The authors used the information that the degradation in tracking performance has distinct patterns dependent on payload variation. The multilayer perceptrons are trained off-line on a representative set of data to recognize the payload variation that produced the tracking error patterns. The training exemplars represent the vectors containing desired and actual position information. The appropriate payload classes are included in each training exemplar to indicate the proper output for the position information. A back propagation algorithm is applied to train the multilayer perceptron to detect and identify each proper class. During the on-line robot control the neural network uses the time frame and information on the desired and actual position to estimate the payload class. This payload class information is needed to recompute the inertial parameters used by the robot's feedforward compensator. The authors call this integration of the principles of connectionist approach and an model-based control "adaptive model-based neural network controller".

The connectionist approach is very efficient in the case of robots with flexible links, or for a flexible materials handling system by a robotic manipulator, where the parameters are not exactly known and a learning capability is important to deal with such problems [109], [143]. Because of the complex nonlinear dynamic model, the recurrent neural network is very convenient for the compensation of flexible effects. In [109], one practical implementation of a multiloop connectionist controller for a single flexible link was realized and its performance compared to standard PD and PID controllers. The connectionist controller consists of an outer PD tracking loop, a singular perturbation inner loop for stabilization of the fast dynamics and a connectionist inner loop, used to feedback linearize the slow system dynamics. This controller requires no off-line learning phase, while experiments showed that standard PD and PID controllers are not able to track a varying desired trajectory. However, the connectionist controller takes the tracking error very close to zero, guaranteeing the boundedness of the tracking error and control signal.

In the area of application of the connectionist approach for mobile robots, real-time fine motion control of a nonholonomic mobile robot is investigated in [133], where both the robot dynamics and geometric parameters are completely unknown. A novel neural network controller combining both kinematic control and dynamic control is developed. The neural network assumes a single layer structure by taking advantage of robot regressor dynamics that express the highly nonlinear robot dynamics in a linear form, in terms of the known and unknown robot parameters. The learning algorithm is computationally efficient. The system stability and the convergence of tracking errors to zero are rigorously proved using the Lyapunov stability theory. The real-time fine control of mobile robot is achieved through the on-line learning of the neural network. In addition, the developed controller is capable of learning the kinematic parameters on-line. The effectiveness and efficiency of the proposed controller is demonstrated by simulation studies.

In [224], a framework for the integrated planning and control for mobile robot navigation is presented. This integrated framework consists of four levels. At the highest level, the planning module produces a sequence of checkpoints from the starting point to the goal, using a variation of the cell decomposition method. The main difference is that our algorithm operates in the robot's workspace instead of the configuration's space. The motion path between checkpoints is determined by the target-reaching module. It senses the checkpoint state relative to the current state and outputs appropriate motor control signals. For mobile robot navigation, this module contains a neural network that is trained to produce a sequence of low-level (motor velocity) control commands to move the robot from one checkpoint to the next. The next lower-level module, obstacle avoidance, senses the presence of local, unforeseen or moving obstacles and produces additional motor control commands to impel the robot away from obstacles. The lowest-level homeostatic control module senses the internal states of the robot in order to maintain internal stability by coordinated responses that automatically compensate for environmental changes. This module is not strictly required for mobile robot navigation but is crucial for mobile manipulation tasks, where the robot is manipulating an object or the environment while its base is in motion. The three lower-level modules constitute a reactive model of motion control. The command fusion module combines the control commands from the reactive components into a final command that is sent to the actuators. All the modules operate asynchronously, at different rates. The integrated framework is applicable to both mobile robot and robot manipulator. This paper focuses on the two lower-level modules (target reaching and obstacle avoidance) and their integration with the planning module, with specific application to target reaching by a nonholonomic mobile robot. This task is performed by an extended Kohonen map (EKM), which is trained to produce a sequence of motor velocity commands. Quantitative experimental results show that the neural network can perform fine control of the motion of a mobile robot very accurately and efficiently. In addition, qualitative test results show that the low-level reactive control modules can be seamlessly integrated with the high-level planning module. In particular, changes to the robot's heading can be easily made at every level even when the robot is en route to the goal position. Syam et al. [328] proposed a new adaptive actor-critic algorithm for the control of mobile robots under the assumption that a predictive model (such as a kinematic model of robot) is available and only the measurement taken at time k is used to update the learning algorithms. Two value-functions are realized as a pure static mapping, according to the fact that they can be reduced to nonlinear current estimators, which can be easily constructed by using any of the artificial neural networks with sigmoidal function or radial basis function, if all the inputs to the present value functions are based on simulated experiences generated from the predictive model. In addition, if a predictive model is assumed to be used to construct a model-based actor (MBA) in the framework of the adaptive actor-critic approach, then this type of MBA can be viewed as a network whose connection weights are composed of the elements of the feedback gain matrix, so that temporal difference learning can also be naturally applied to update the weights of the actor. Since the present method can update the learning by using only one measurement at time k, relatively fast learning is expected, compared with the previous approach that needs two measurements at times k and $k + 1$ to update the actor-critic networks. The effectiveness of the proposed approach is illustrated by simulating a trajectory-tracking control problem for a nonholonomic mobile robot.

The connectionist systems also appeared to have a very useful role in the dynamic control of underwater robot manipulators (autonomous undewater vehicles - AUV) [207], [206] [140], citelor96, [395], [396], [214]. Unlike industrial robot manipulators, the dynamics of underwater robot manipulators have large uncertainties owing to buoyancy, the force induced by the added mass/moment of inertia and the drag force together with various disturbances caused by sea flow, waves and currents. In [207], the authors have proposed a robust controller, which consists of conventional computed torque control, sliding mode controller and back-propagation neural network. The on-line training of the neural controller enables successful compensation for the control error of sliding mode controller with an unsuitable initial assumption of uncertain bounds. Yuh [395],[396] proposed a neural network control system using a recursive reinforcment learning algorithm. A special feature of this controller is that the robotic system adjusts itself directly and on-line without an explicit model of underwater vehicle dynamics. Lorenz and Yuh [223] presented experimental results on Omni-Directional Intelligent Navigator (ODIN) by the previously mentioned connectionist method. Ishi et al. [140] have proposed the Self-Organizing Neural-Network Controller System (SONCS) System for AUV's and examined its effectiveness through application to the heading - keeping control of a special AUV. In this research, quick adaptation method called "imaginary training" is used to improve the time-consuming adaptation process of SONCS. In [214], a linearly parameterized neural network (LPNN) is used to approximate the uncertainties of the vehicle dynamics, where the basis function vector of the network is constructed according to the vehicle's physical properties. The proposed controller guarantees the uniform boundedness of the vehicle's trajectory tracking errors and the network's weights estimation errors based on the Lyapunov stability theory, where the network's reconstruction errors and the disturbances in the vehicle's dynamics are bounded by an unknown constant. Numerical simulation studies are performed to illustrate the effectiveness of the proposed control scheme. From these simulation results, we can see that the constructed neural network has a satisfactory approximation capacity under the above simulation conditions and weights update laws, which clearly resulted in superior tracking performance.

Various types of neural networks are used for the control design of humanoid robots, such as multilayer perceptrons, CMAC (Cerebellar Model Arithmetic Controller) networks, recurrent neural network, a RBF networks or Hopfield networks, which are trained by supervised or unsupervised learning methods. The majority of the proposed control algorithms have been verified by simulation, while there have been few experimental verifications on real biped and humanoid robots. Neural networks have been used as efficient tools for solving static and dynamic balance during the process of walking and running on terrain with different environmental characteristics.

Kitamura et al. [187] proposed a walking controller based on the Hopfield neural network in combination with an inverted pendulum dynamic model. The optimization function of the Hopfield network was based on a complete dynamic model of the biped.

Wang et al.[366] have developed a hierarchical controller for a three-link two-legged robot. The approach uses the equations of motion, but only for the training of the neural networks rather than to directly control the robot. The authors used a very simplified model of a biped with decoupled frontal and sagittal planes. There are three neural networks (multilayer perceptrons) for control of the leg on the ground, control of the leg in the air, and for body regulation. This approach uses off-line training and on-line adaptation. The training algorithm is a standard back-propagation algorithm based on the difference between the decoupled supervising control law and output of all three neural networks. There is no feedback in real-time control, and this is a great problem in cases where system uncertainties exist.

Apart from considering the walking control problem, very little research has been done on the problem of intelligent control of running. Doerschuk et al. [66] presented an adaptive controller to control the movement of simulated jointed leg during a running stride (uniped control). The main idea of this approach is the application of modularity, i.e., the use of separate controllers for each phase of the running stride (take-off, ballistic, landing), thus allowing each to be optimized for the specific objective of its phase. In the take-off phase, the controller's objective is to realize inverse feedforward control (for desired height, distance and angular momentum it is necessary to produce control signals that achieve these objectives). The controller learns from experience to produce the control signals that will produce the desired height, distance and angular momentum. Three different types of neural networks are investigated (multilayer perceptrons, CMAC, and neuro-fuzzy nets). It was concluded that neuro-fuzzy nets achieve more accurate results than the other two methods. Off-line global training is not needed because of the use of local learning. The neuro-fuzzy take-off controller controls very accurately the value of angular momentum of the stride after only two learning iterations. The ballistic controller controls the movement of the leg while the foot is in the air. In this case, ballistic controller combines neural network learning with the conventional PD control. It is a typical feedback error learning scheme, where a PD controller generates the torques that are applied to the joints, producing movement of the leg. The controller learns the dynamic model of the leg from experience generated by the PD controller and improve upon its performance. The CMAC controller is used for the neural network learning part with the possibility to very accurately control the movement of the leg along a target trajectory, even at the first attempt. Ballistic learning is accomplished on-line without the need for precomputed examples. This enables the effective adaptability of the humanoid robot to various changes and new conditions.

Much work has been done by researchers in order to present stable connectionist control algorithms for robot trajectory tracking [212], [291], [154], [149], [314], [55], [185], [211]. There are some common features in these algorithms. The first is the use of locally generalizing networks for fast learning and adaptation (Gaussian RBF nets, CMAC nets or basis-spline nets). The other characteristic is the use of the Lyapunov stability theory or passive theory to provide the global stability of the system. Mathematical proofs of stability lift the neural network control design to the level of firm engineering theory and practice.

4.3   Sensor-based robot learning by neural networks

A completely different approach is represented by the connectionist method, which uses sensory information for robot neural control. Sensor-based control is a very efficient method of overcoming the problems with robot model and environment uncertainties, because sensor capabilities help in the adaptation process without explicit control intervention. Specifically, it is an adaptive sensor-motor coordination that uses various mappings given by the robot sensor system. Particular attention has been paid to the problem of visuo-motor coordination [249],[250],[196],[235] ,[113], [324], [73], [45], [188], [27], [83], in particular for eye-head and arm-eye systems. In general, in visuo-motor coordination by neural networks (perceptrons, neocognitron, CMAC, ARTMAP networks or Kohonen maps) visual images of the mechanical parts of the system can be directly related to posture signals. A general type of application involves locating a target object so that it can be grasped by a robotic manipulator. The connectionist system was used to transform the image of an object into the object's pose relative to the manipulator, which is passed on to a conventional robot controller for the further model-based kinematic transformations and low-level motor-control functions necessary to complete the object capture.

After many years of being referred to as impractical in robot control, it was demonstrated that CMAC could be very useful in the learning of state space-dependent control responses for visuo-motor coordination [249], [250]. A typical demonstration of the application of CMAC in robot control involves the control of an industrial robot using a video camera. The robot's task is to grasp an arbitrary object lying on a conveyor belt with fixed orientation, or to avoid various obstacles in the working space. In the learning phase, visual input signals about the objects are processed and combined into a target map through modifiable weights, which generates the control signals for the robot motors. The differences between the actual motor signals and the motor signals computed from the camera input are used to incrementally change the weights. After the learning phase, in the pattern association phase, the learned sensory-motor correlation is used to recognize and manipulate the objects, which is similar to "training" objects. This neural architecture consists of motor map representations interleaved within sensory topography. It is not a model-based approach. Hence, there is no need for a kinematic robot model. Thus, this method can be used for any kind of system, as long as there are enough units and random connections. Possible drawbacks of the CMAC approach are poor generalization properties and noise due to hash coding.

Kuperstein [196] has presented a similar approach using the principle of *sensory-motor circular reaction* (Figure 2.11). This method relies on the self-consistency between sensory and motor signals to achieve unsupervised learning. In this scheme, two phases of functioning can be distinguished. During the first phase (learning phase), sensor/motors relations are learnt via correlation between input and output signals. The manipulation robot receives a large, randomly-generated set of postures and orientations, and moves accordingly. At the same time, visual feedback based on the current state of the network represents the second set of commands. The difference between these two sets of commands is used to modify the weights of the network. This incremental learning is based on Hebb's rule. In the second phase, the system uses the learned correlations to evoke the correct posture that manipulates a sensed object. In this case, the random generator and learning rule have to be switched off and the target map is fed directly to the robot actuators. Moreover, this learning scheme requires only the availability of the manipulator, and no formal knowledge of robot kinematics is needed.

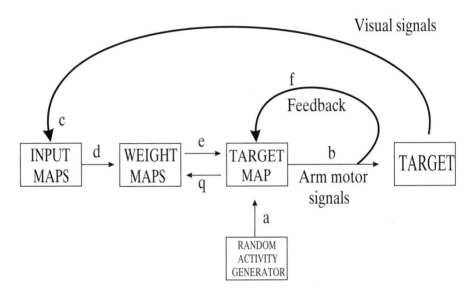

Figure 2.11. Sensor-motor circular reaction

Martinetz, Ritter and Schulten [235] presented an extension of the Kuperstain method with the aim to overcome some problems inherent to this method. They used Kohonen's algorithm [190] for the formation of topologically correct feature maps together with the Widrow-Hoff error-correction rule. This unsupervised approach learns the functional relationship between the input Cartesian space and output (joint) space based on a localized adaptation of the mapping, by using the manipulator itself under joint control and adapting the solution based on the comparison of the resulting location of the manipulator end effector in Cartesian space with the desired location. Instead of using a topographic map which is only one-dimensional, as did Kuperstain, the authors proposed a single three-dimensional map. In this way they eliminated the restriction arising from the additive coupling of several one-dimensional maps. Also, adaptive topographic ordering of map is proposed, which allows the map not only to adapt its outputs, but also to adapt the range of its input signals. In addition, this topology enables many neighbouring units to cooperate during the learning process, which greatly contributes to the efficiency and robustness of the algorithm. This algorithm also uses local representation of the mapping as one of the main features for fast convergence properties. However, on the other hand, this approach has a drawback because of the complexity of adjusting the weights of a great number of neural units. Hence, it is suitable only for low-dimensional problems.

A main drawback of these neural approaches is the long training period. The "black box" approach results in increased learning time and hinders the real-time implementation of practical tasks. Hence, in [27] Behera and Kirubanandan suggest an approach using Kohonen SOFM to learn the hand-eye coordination problem in reduced time but with high accuracy. This hybrid method modifies previous neural approaches by utilizing a model of the system for initial learning and on-line (model-free) training on the system and continues the learning process during the operational phase, thereby adapting to the system changes as they occur. The hybrid scheme achieves high accuracy in the entire workspace, improving on the model-based approach and taking much less time than the model-free approach. Experimantal results show that the proposed neural scheme is on an average 10 times faster in training, compared to similar neural approaches.

In [83], a vision-based motion planning method is presented. The method uses the Topology Representing Network (TRN) to learn the topology representation of the Perceptual Control Manifold (PCM) for the end-effector of the robot arm. Also, a way of mapping obstacles from the 3-D work space on to the learned representation using the connections between neurons, is proposed. The results of the simulation and experiment with robots PUMA 562 and Mitsubishi RV-E2M demonstrate the feasibility of the approach.

Visual servoing in robotics considers the mapping of input geometric features into the output robot velocity by inverse feature Jacobian. Feature selection and inverse Jacobian derivation can be very tedious, hence many researchers have used various types of neural networks (back propagation networks, SOFM) to automate and approximate the inverse feature Jacobian and feature extraction [317], [40], [302].

The tactile-motor coordination differs significantly from the visuo-motor one, due to the intrinsic dependency on the contacted surface. The direct association of tactile sensations with positioning of the robot end-effector cannot be feasible in many cases, hence it is very important to understand how a given contact condition will be modified by motor action. In this case the task of the neural network is to estimate the direction of the feature-enhancement motor action on the basis of modifications in the sensed tactile perception. In contrast to the previously mentioned approaches to visuo-motor coordination, in [284] the authors developed and experimentally verified the autonomous learning of tactile-motor coordination by Gaussian network for simple robotic system composed of a single finger mounted on the robot's arm.

Some more recent and sophisticated learning architectures (adaptive learning architecture) involve a neural estimator that identifies some robot parameters using available information from robot sensors (Figure 2.12). Based on information from the neural estimator, the robot controller modifies its parameters and then generates a control signal for the robot actuators. The robot sensors observe the status of the system and make available information and parameters to the estimator and robot controller. Based on this input, the neural estimator changes its state, moving in the state space of its variables. The state variables of the neural estimator correspond exactly to the parameters of the robot controller. Hence, the stable-state topology of this space can be designed so that the local minima correspond to an optimal law.

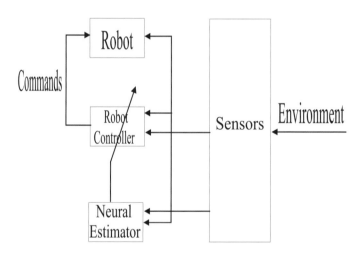

Figure 2.12.   Sensor-based learning architecture

Neural networks can be efficiently employed as pattern classifiers in robotic applications. For example, an interesting application of neural networks for processing sonar signals for target differentiation and localization for robot indoor environments was presented in [24]. In this case a simple neural network is capable of making an automatic selection from the discriminating functions. These functions offer the ability to distinguish a larger number of target types than standard methods, often with greater accuracy. In addition, the neural network for this type of application is found to be robust to a variety of failure sensing modes. The presented results suggest a wider use of neural network

as robust pattern classifiers in sensor-based robotics. An interesting approach
to fault detection and indentification in a mobile robot using Kalman Filter
estimation and neural network is presented in [96]. The aim of this research
was to show that it is possible to detect and identify both sensor and mechanical
failures on a mobile platform by means of analyzing the collective output of a
bank of Kalman Filters. The proposed neural network, trained with the back
propagation method, is used for processing the set of filter residuals given by
the filters as a pattern and deciding which fault has ocurred, that is, which filter
is most reliable.

Salatian et al. [289], [290], [288] studied off-line and on-line reinforcement
techniques for adapting the gait designed for horizontal surfaces to sloping surfaces. They considered the humanoid robot SD-2 with 8 DOFs and two force
sensors on both feet. These control algorithms, without considering kinematic
and dynamic models of humanoid robot, were evaluated using a biped dynamic
simulation [289], [290], as well as on the real biped SD-2 [288]. The control
structure includes a gait trajectory synthesizer (with a memory of previously
stored and learned gaits) and neural networks that are tuned by reinforcement
signals from force sensors at the feet. The joint positions of the robot are adjusted until the force sensors indicate that the robot has a stable gait. The neural
network has the task to map the relation between foot forces and to adjust the
joint positions. Reinforcement learning is used because the neural network
receives no direct instruction on which joint position needs to be modified.
The neural network is not a conventional type of network (perceptrons) and
includes a net of more neurons with inhibitory/excitory inputs from the sensor
unit. Every joint of the robot is associated with a neuron called a *joint neuron*.
Every joint neuron is further attached to two pairs of neurons, called *direction
neurons*. Each neuron possesses a value of activation function called *neuron
value*. During the learning process, a joint neuron with the maximum neuron
value is selected to modify the position of the corresponding joint, while a direction neuron is selected to determine the direction of modification. If the selected
joint and direction neuron result in a correct motion (the robot becomes more
stable), the selection is reinforced by increasing the neuron value. Otherwise,
the neuron value is reduced. Using the previously mentioned "reward-and-
punish" strategy, the neural network converges quickly and generates a stable
gait on the sloping surface. In this way, reinforcement learning is very attractive
because the algorithm does not require an explicit feedback signal. The computation issues for reinforcement learning are simple, while the noise from the
feet sensors is taken into the process of learning. During one step of the biped,
there are eight static configurations, called the primitive points. The neural
network is only responsible for motion in the sagittal plane. Hence, including
the redundancy at the hip joint. there are 3 x 8 = 24 joint neurons. Because
of the nature of reinforcement learning, only one joint neuron is active each

time. Static and pseudo-dynamic are demonstrated to prove that the proposed mechanism is valid for robot walking on a sloping surface. In this approach, kinematic and dynamic models were not used, hence, it would be a problem for real dynamic walking at high speed. Also, real terrain is more complex than the environments used in test experiments, so that more studies need to be conducted to make the robot walk robustly on different sorts of terrain.

More recently, Miller [247],[248],[194] have developed a hierarchical controller that combines simple gait oscillators, classical feedback control techniques and neural network learning, and does not require detailed equations of the dynamics of walking. The emphasis is on the real-time control studies using an experimental ten-axis biped robot with foot force sensors. The neural network learning is achieved using a CMAC controller, where CMAC neural networks were used essentially as context-sensitive integral errors in the controller, the control context being defined by the CMAC input vector. There are three different CMAC neural networks for humanoid posture control. The Front/Back Balance CMAC neural network was used to provide front/back balance during standing, swaying and walking. The training of this network is realized using data from foot sensors. The second CMAC neural network is used for Right/Left Balance, to predict the correct knee extension required to achieve sufficient lateral momentum for lifting the corresponding foot for the desired length of time. The training of this network is realized using the temporal difference method, based on the difference between the desired and real time of foot rising. The third CMAC network is used to learn kinematically consistent robot postures. In this case, training is also realized by data from the foot sensors.

The results indicated that the experimental biped was able to learn the closed-chain kinematics necessary to shift body weight from side to side while maintaining good foot contact. It was also able to learn the quasi-static balance required to avoid falling forward or backward while shifting body weight side-to-side at different speeds. It was able to learn the dynamic balance in order to lift a foot off the floor for a desired length of time and different initial conditions. There were, however, many limitations (limited step length, slow walking, no adaptation for left-right balance, no possibility of walking on sloping surfaces). Hence upgrading and improvement of this approach were proposed in [194]. A new dynamically balanced scheme for handling variable-speed gait was proposed based on preplanned but adaptive motion sequences in combination with closed-loop reactive control. This allows the algorithm to improve the walking performance over consecutive steps using adaptation, and to react to small errors and disturbances using reactive control. New sensors (piezoresistive accelerometers and two solid-state rate gyroscopes) are mounted on the new UNH biped (Figure 2.13). The complete control structure consists of high-level and low-level controllers (Figure 2.14). The control structure of high-level con-

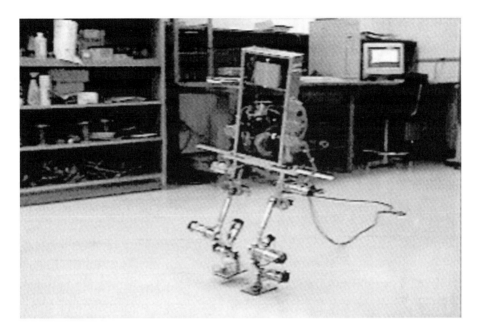

Figure 2.13. The UNH biped walking

trol includes seven components (Figure 2.15): gait generator, simple kinematics block and 5 CMAC controllers. The following posture parameters are presented in Figure 2.15: H - height; Rff and Lff are right and left foot forward; Rfl and Lfl are right and left foot lift; Fbl is front-back lean and Rll is right-left lean. The operation of the gait generator is based on simple heuristics and an appropriate biped model. The CMAC neural networks are used for the compensation of right and left lift-lean angle correction, reactive front-back offset, right-left lean correction, right and left ankle - Y correction and front-back lean correction. Training of the neural networks is realized through the process of temporal difference learning, using information about zero-moment point (ZMP) from robot foot sensors. The five CMAC neural networks were first trained during repetitive foot-lift motions, similar to marching in place. Then, training was carried out during the attempts at walking for increased step length and gait speeds. The control structure on the lower control level includes reactive lean angle control, together with a PID controller.

The experimental results indicate that the UNH biped robot can walk with forward velocities in the range of 21 - 72 cm/min, with sideways leaning speed in the range of 3.6 - 12.5 o/s. The main characteristic of this controller is the synthesis of the control signal without a dynamic model of the biped. The proposed controller could be used as a basis for similar controllers of more

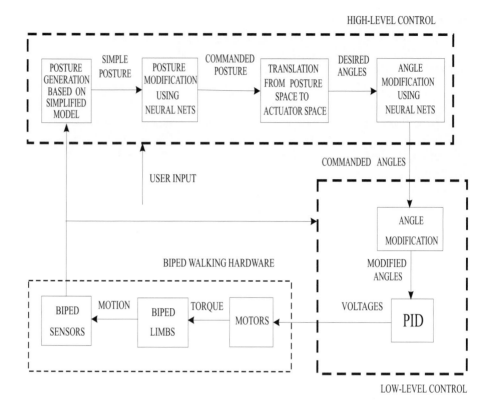

Figure 2.14. Block diagram of overall biped control system

complex humanoid robots in future research. However, this controller is not of a general nature, because it is suitable only for the proposed structure of biped robot and must be adapted to bipeds with different structures. More research efforts are needed to simplify the controller structure, to increase the gait speed and ensure the stability of dynamic walking.

The previously used CMAC controller is a particularly good option for robotic motor control. It has the quality of fast learning and simple computation in comparison with multilayer perceptrons and similar approximation capabilities such as RBF networks. However, there are problems with large memory requirements, function approximation and stability of dynamic walking. These problems have been addressed in [131], where a self-organizing CMAC neural networkstructure was proposed for biped control based on a data clustering technique together with adaptation of the basic control algorithm. In this case, the memory requirements are drastically reduced and globally asymptotic stability is achieved in a Lyapunov sense. The structural adaptation of the network centres is realized to ensure adaptation to unexpected dynamics. Unsupervised learning using CMAC can be implemented with a Lyapunov trajectory index. The distance between the input vector and the centre vectors of the CMAC is calculated, then the memory cells corresponding to the centres (hit by the input) are found, and finally, computation of the CMAC output by a linear combination of CMAC basis functions and weights of the memory cell is achieved. The weights in the fired memory cells are updated by unsupervised learning. The approach is verified through simulation experiments on a biped with seven DOFs. An important characteristic of this approach is the inclusion of adaptation for CMAC and PID controllers with a moderate increase of controller complexity to handle disturbances and environmental changes. Although the robustness was enhanced in terms of height and pitch tracking along with external disturbance rejection, the proposed adaptive controller does not guarantee long-term stability of the walking gait. The following posture pa-

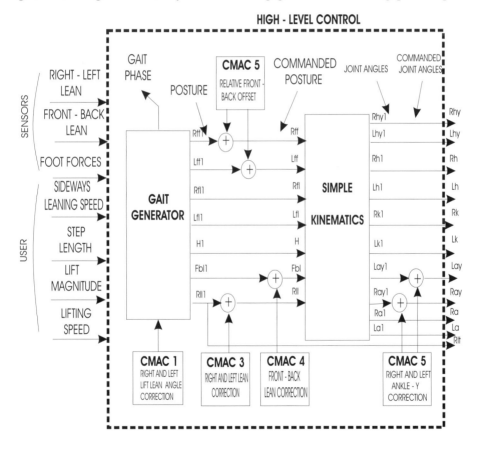

Figure 2.15. High - level control architecture

rameters are presented in Figure 2.15: H - height; Rff and Lff are right and left foot forward; Rfl and Lfl are right and left foot lift; Fbl is front-back lean and Rll is right-left lean.

5.   Efficient Learning Control for Manipulation Robots by Feedforward Neural Networks

In this section we shall present some new robot control learning algorithms with fast and robust learning properties using several different network architectures and learning schemes. Our major concern is the application of neural networks in robot control on the executive hierarchical level for learning inverse dynamic model of the robot mechanism. The synthesis of new learning control algorithms with fast learning rules based on multilayer perceptrons is performed. Finally, the proposed hybrid learning approach, in addition to fast neural network algorithms, uses the deterministic a priori known part of the dynamic model in an analytical form. In this way, the process of training neural network is accelerated. Another important feature of the proposed learning control structures is fast convergence properties, because the problem of adjusting the weights of the internal hidden units is considered as a problem of estimating parameters by recursive least squares (RLS) or Extended Kalman Filter (EKF) method. In this way, special new algorithms with time-varying learning rate can yield benefits to learning speed and generalization.

5.1   Efficient connectionist learning structures

One of the most important problems in solving robot control synthesis is the high nonlinearity of the robot dynamics model with expressive couplings between robot subsystems. The dynamic model of a manipulation robot can be written as

$$P = f(q, \dot{q}, \ddot{q}, \theta) = H(q, \theta)\ddot{q} + h(q, \dot{q}, \theta) \qquad (2.39)$$

where $P \epsilon R^n$ is the vector of driving torques or forces; $H(q, \theta) : R^n \times \theta \Rightarrow R^{n \times n}$ is the inertial matrix of the system; $h(q, \dot{q}, \theta) : R^n \times R^n \times \theta \Rightarrow R^n$ is the vector which includes centrifugal, Coriolis and gravitational effects; $\theta \epsilon R^{nt}$ is the system parameter vector; n is the number of DOFs; nt is the number of system parameters.

A common and conventional way for robot control represents local PID regulators for each DOF of the robotic mechanism:

$$u = u_{fb} = -KP \, \varepsilon - KD \, \dot{\varepsilon} - KI \int \varepsilon dt \qquad (2.40)$$

where $u \epsilon R^n$ is the control input; $u_{fb} \epsilon R^n$ is the feedback control; $KP \epsilon R^{n \times n}$ is the matrix of local position feedback gains; $KD \epsilon R^{n \times n}$ is the matrix of local velocity feedback gains; $KI \epsilon R^{n \times n}$ is the matrix of local integral feedback gains; $\varepsilon = q - q_d$ is the feedback error ($\varepsilon \epsilon R^n$); q and q_d are the real and desired internal coordinates ($q \epsilon R^n$, $q_d \epsilon R^n$). However, this control law is inadequate for advanced industrial robots, because the influence of couplings between the subsystems is substantial. Hence, "dynamic control", which takes the dynamic model of the robot mechanism in control synthesis in the form of feedforward control, is included as a solution. In view of the above, a *decentralized control algorithm* is applied:

$$u = u^{ff} - KP\,\varepsilon - KD\,\dot{\varepsilon} - KI \int \varepsilon dt \qquad (2.41)$$

where $u^{ff} \epsilon R^n$ is the nominal centralized or decentralized feedforward control, synthesized off-line using the integral robot model (with the model of the robot actuators). However, in the procedure of controller design, control designers have to cope with structured uncertainties (inaccuracies of model parameters and additive disturbances), unstructured uncertainties (unmodelled high-frequency dynamics as structural resonant modes, neglected time-delays, actuator dynamics, sampling effects, etc.) and measurement noise. The time-varying nature of robot parameters and the variability of robot tasks represent additional difficulties for the control system. In this case, the application of conventional nonlearning adaptive control is not sufficient for high-quality performance. Therefore, a solution of the robot control problem will probably require combining conventional approaches with new learning approaches in order to achieve good performance. The connectionist methods provide the implementation tools for efficient learning of complex input-output relations of robot dynamics and kinematics. One of the main goals of dynamic learning methods is to find a solution to the robot inverse dynamics problem. In this way, a connectionist structure, as part of a decentralized feedforward control law, can compensate for a wide range of robotic uncertainties.

The connectionist structure is commonly used as part of the feedforward controller in a decentralized control algorithm. Training and learning by connectionist structure is accomplished exclusively in an on-line regime by the feedback-error learning method.

The application of multilayer perceptrons for robot learning algorithms is defined here by a four-layer perceptron with *symmetric sigmoid* function as activation function in both hidden layers. The activation function for input and output layers is the identity function. The number of neurons in the hidden layers ($12n + 1$ neurons in the first hidden layer, $6n + 1$ neurons in the second hidden layer) is determined by simulation experiments and experience. The neural network with the proposed topology of fixed nonrecurrent multilayer network generates the necessary driving torques at robot joints as a nonlinear mapping:

$$P_i^{NN} = g(w_{jk}^{ab}, q_d, \dot{q}_d, \ddot{q}_d) \qquad i = 1, .., n \qquad (2.42)$$

where $P_i \epsilon R^n$ is the joint driving torque generated by the neural network; w_{jk}^{ab} are the adaptive weighting factors between the neuron j in the $a - th$ layer and the neuron k in the $b - th$ layer; g is the nonlinear mapping.

The presented "black-box" approach can yield good results in robot learning control, but convergence of the learning process may in some cases be very slow because the algorithm does not use available deterministic information about the robot model structure. Hence, an interesting idea of using a hybrid learning approach has been proposed [167]. This approach is based on using simultaneously the "known" deterministic part of inverse robot dynamics in analytic form and multilayer perceptrons for learning and compensation of system uncertainties, which are not captured by the deterministic robot model. Using the topology of a fixed four-layer nonrecurrent perceptron, part of the controlled torque for reduction of system uncertainties is given by the following relation:

$$P_i^{NNkom} = g_r(w_{jk}^{ab}, q_d, \dot{q}_d, \ddot{q}_d) \quad i = 1, ., n \tag{2.43}$$

where $P_i^{NNkom} \epsilon R^n$ is the part of the controlled torque; g_r is the nonlinear mapping by the neural network.

The controlled torque, based on the deterministic dynamic model of rigid robot links and desired robot trajectory, is given in the following form:

$$P_i^{kr} = \sum_{j=1}^{n} H_{ij}(q_d, \theta)\ddot{q}_d + h_i(q_d, \dot{q}_d, \theta) \tag{2.44}$$

Using the integral robot model and the previous two equations, the decentralized control algorithm based on the hybrid approach has the following form:

$$u_i = u_i^{ff}(q_d, \dot{q}_d, \ddot{q}_d, P_i^{kr} + P_i^{NNkom}) + u_i^{fb} \tag{2.45}$$

The back propagation algorithm caused a significant breakthrough in the control application of multilayer perceptrons. On the other hand, the standard BP-learning algorithm as a basic training algorithm for multilayer perceptrons suffers from the typical handicaps of all steepest descent approaches: very slow convergence rate and the need for predetermined learning parameters limit the practical use of this algorithm. Here, we will consider the problem of adjusting the weights of internal hidden units as a problem of parameter estimation by well-known estimation methods (Recursive Least Square method and Extended Kalman Filter approach) [167],[164]. The use of these methods with time-varying learning rates yields benefits to learning speed and generalization that are not available with the standard back-propagation algorithm. The proposed new algorithms are based on previously defined 4-layer connectionist structures with appropriately defined parameters of the network. The main relations in the process of training for forward pass in the network are described by the following expressions:

$$s^2(k) = W^{12}(k)^T i^1(k) \qquad (2.46)$$

$$o_a^2(k) = 1/(1 + exp(-s_a^2(k))) \qquad a = 1, ..., L_1 \quad o_0^2(k) = 1 \quad (2.47)$$

$$s^3(k) = W^{23}(k)^T o^2(k) \qquad (2.48)$$

$$o_b^3(k) = 1/(1 + exp(-s_b^3(k))) \qquad b = 1, ..., L_2 \quad o_0^3(k) = 1 \quad (2.49)$$

$$s^4(k) = W^{34}(k)^T o^3(k) \qquad (2.50)$$

$$y_c(k) = s_c^4(k) = P_c^{NN}(k) \quad c = 1, .., n \qquad (2.51)$$

where $s^2(k), s^3(k), s^4(k)$ are the output vectors for linear parts of the layers; $o^2(k)$, $o^3(k)$ are the output vectors of the hidden layers; $W^{12} = [w_{m \times L1}^{12}]$, $W^{23} = [w_{L1+1 \times L2}^{23}]$, $W^{34} = [w_{L2+1 \times n}^{34}]$ are the weighting factors of the layers; $i^1(k)$ is the input to the network (robot internal positions, velocities and accelerations - $m = 3n + 1$); $y(k)$ is the network output.

A first approach applied to estimate weights at the hidden layers was the RLS method for parameter estimation of linear systems. The general idea is that multilayer perceptrons can be observed as a set of sequential linear decomposed subsystems, which are connected by nonlinear connections. As is well known, recursive estimators for linear deterministic systems show faster convergence properties than gradient estimators (BP algorithm). Hence, the aim of estimation is to define optimal values for the matrices W^{12}, W^{23}, W^{34} using models of linear systems. In the application of this method for specification of desired values of the linear parts of layers $s^{id}(i = 2, 3)$, two different techniques are used: gradient approximation and inverse mapping. The basic equations describing new learning rules based on the RLS gradient approximation method are given by the following formulae [167]:

$$for \quad h = 4$$

$$s_c^{hd}(k) = y_c^d(k) = P_c^d(k) \qquad c = 1, .., n \qquad (2.52)$$

$$for \quad h = 3, 2 \qquad g = 4, 3$$

$$s_t^{hd}(k) = s_t^h(k) + \alpha_t \delta^g(k)^T W^{hg}(k)^T o^h(k)[1 - o_t^h(k)]$$
$$t = 1, ., L_2 \quad or \quad t = 1, ., L_1 \qquad (2.53)$$

$$for \quad h = 3, 2, 1$$

$$C^h(k) = \frac{1}{\lambda^h} \{ C^h(k-1) - \frac{C^h(k-1)o^h(k)o^h(k)^T C^h(k-1)}{\lambda^h + o^h(k)^T C^h(k-1)o^h(k)} \} \quad (2.54)$$

$$for \quad h = 3, 2, 1 \qquad g = 4, 3, 2$$

$$W^{hg}(k) = W^{hg}(k-1) + C^h(k)o^h(k)[s^{gd} - W^{hg}(k-1)^T o^h(k)]^T \tag{2.55}$$

$$for \quad h = 4, 3 \qquad g = 3, 2$$

$$\delta_t^h(k) = s_t^{hd}(k) - \sum_j w_{tj}^{gh}(k)o_j^g(k)$$

$$t = 1, .., n \quad j = 1, .., L_2 + 1 \quad or$$

$$t = 1, .., L_2 \quad j = 1, .., L_1 + 1 \tag{2.56}$$

where λ^h and α_t are the appropriate forgetting factors and learning rates; C^h is the appropriate covariance matrix.

In the second case, connections between the linear and nonlinear parts of the neural network are achieved by inverse mapping of the activation function in the network hidden layers [168]. In this case, equation (2.53) is replaced by the following equations:

$$for \quad h = 3, 2 \quad g = 4, 3$$

$$o^{hd}(k) = o^h(k) + W^{hg}(k)\delta^g(k) \tag{2.57}$$

$$o_{max}^{hd}(k) = max \mid o_t^{hd}(k) \mid \tag{2.58}$$

$$o_{min}^{hd}(k) = min \mid o_t^{hd}(k) \mid \tag{2.59}$$

$$o_t^{hp}(k) = 0.01 + (o_t^{hd}(k) - o_{min}^{hd}(k))\frac{0.98}{o_{max}^{hd}(k) - o_{min}^{hd}(k)} \tag{2.60}$$

$$s_t^{hd}(k) = -ln\frac{1 - o_t^{hp}(k)}{o_t^{hp}(k)} \qquad t = 1, .., L_2 \quad or \quad t = 1, .., L_1 \tag{2.61}$$

In a similar way, considering the whole network as a set of nonlinear decomposed subsystems, the EKF approach to estimation of weighting factors is applied. In this approach, an optimal estimate of system parameters is calculated for each input pattern with fast convergence properties and without tuning parameters. Also, in order to reduce numerical complexity, weighting matrices are updated in the loop as column vectors. For example, the learning algorithm via the EKF gradient approximation approach is defined according to the following expressions [164]:

$$\delta_c^4(k) = P_c^d(k) - P_c^{NN}(k) \quad c = 1, .., n \tag{2.62}$$

Loop according to i from 1 to n

$$\hat{\Delta}^{34i}(k) = [0, \cdots, \overset{i}{1}, 0, \cdots, 0]^T \qquad (2.63)$$

$$H_i^{34}(k) = \hat{\Delta}^{34i}(k)o^{3^T}(k) \qquad (2.64)$$

$$K_i^{34}(k) = C_i^{34}(k-1)H_i^{34^T}(k)[H_i^{34}(k)C_i^{34}(k-1)H_i^{34^T}(k)+r_2^{34}(k)I]^{-1} \qquad (2.65)$$

$$W_i^{34}(k) = W_i^{34}(k-1) + K_i^{34}(k)\delta^4(k) \qquad (2.66)$$

$$C_i^{34}(k) = C_i^{34}(k-1) - K_i^{34}(k)H_i^{34}(k)C_i^{34}(k-1) + r_1^{34}(k)I \qquad (2.67)$$

End of the loop according to i

$$o^{3z}(k) = o^3(k) + W_{34}(k)\delta^4(k) \qquad (2.68)$$

$$s_b^{3z}(k) = s_b^3(k) + \alpha^3 o_b^3(k)(1 - o_b^3(k))[o_b^{3z}(k) - o_b^3(k)] \qquad b = 1, .., L_2 \qquad (2.69)$$

Loop according to i from 1 to L_2

$$\hat{\Delta}^{23i}(k) = o_i^3(k)[1 - o_i^3(k)][0, \cdots, \overset{i}{1}, 0, \cdots, 0]^T \qquad (2.70)$$

$$H_i^{23}(k) = \hat{\Delta}^{23i}(k)o^{2^T}(k) \qquad (2.71)$$

$$K_i^{23}(k) = C_i^{23}(k-1)H_i^{23^T}(k)[H_i^{23}(k)C_i^{23}(k-1)H_i^{23^T}(t)+r_2^{23}(k)I]^{-1} \qquad (2.72)$$

$$W_i^{23}(k) = W_i^{23}(k-1) + K_i^{23}(k)[o_b^{3z}(k) - o_b^3(k)] \qquad (2.73)$$

$$C_i^{23}(k) = C_i^{23}(k-1) - K_i^{23}(k)H_i^{23}(k)C_i^{23}(k-1) + r_1^{23}(t)I \qquad (2.74)$$

End of the loop according to i

$$o^{2z}(k) = o^2(k) + W_{23}(k)[s^{3z}(k) - W_{23}(k)^T o^2(k)] \qquad (2.75)$$

Loop according to i from 1 to L_1

$$\hat{\Delta}^{12i}(k) = o^{2i}(k)(1 - o^{2i}(k))[0, \cdots, \overset{i}{1}, 0, \cdots, 0]^T \qquad (2.76)$$

$$H_i^{12}(k) = \hat{\Delta}^{12i}(k)i^{1^T}(k) \qquad (2.77)$$

$$K_i^{12}(k) = C_i^{12}(k-1)H_i^{12^T}(k)[H_i^{12}(k)C_i^{12}(k-1)H_i^{12^T}(k)+r_2^{12}(k)I]^{-1} \qquad (2.78)$$

$$W_i^{12}(k) = W_i^{12}(k-1) + K_i^{12}(k)[o^{2z}(k) - o^2(k)] \qquad (2.79)$$

$$C_i^{12}(k) = C_i^{12}(k-1) - K_i^{12}(k)H_i^{12}(k)C_i^{12}(k-1) + r_1^{12}(k)I \qquad (2.80)$$

End of the loop according to i

where r_1^{ij}, r_2^{ij}, are the variances for zero-mean white Gaussian noise.

5.2   Case Study

In this section we describe simulation trials that were run to verify the proposed connectionist algorithms in compensating the system uncertainties. The manipulation robot used for the simulation was a 6-DOF industrial robot MANUTEC r3. The parameters of the robot mechanism and actuators are listed in Tables 2.1 and 2.2. In the learning phase, the robot training was accomplished

Table 2.1. Parameters of Manutec r3 manipulation mechanism

Mechanism link no.	1	2	3	4	5	6
Mass $[kg]$	50.	56.5	26.4	28.7	5.2	0.01
Length $[m]$	0.67	0.5	0.73	0.01	0.01	0.01
Moment of inertia $J_x [kgm^2]$	0.	2.58	0.279	1.67	1.25	0.
Moment of inertia $J_y [kgm^2]$	0.	2.73	0.413	1.67	1.53	0.
Moment of inertia $J_z [kgm^2]$	1.16	0.64	0.245	0.81	0.81	0.001

using the motion from point A with internal coordinates $q\epsilon(0.; 1.57; 0.; 0.; 0.; 0.)$ to point B: $q\epsilon(0.3; -1.37; 0.3; -0.3; -0.3; 0.3)$. Time duration of the movement was t=1s with trapezoidal velocity profile. The PD feedback control was chosen with the following values of local feedback gains:

$$KP\epsilon(69.; 56.; 43.; 39.; 42.; 15.); \qquad KD\epsilon(2.; 3.; 1.; 0.7; 0.7; 0.3)$$

In the simulation experiments, the model uncertainties are defined by parametric disturbances with approximately 20 % variation from nominal values. In the second part of simulation experiments, the pure connectionist structure (19-73-37-6 topology) was applied as part of the feedforward control. Learning rate, using back propagation algorithm for all layers, was $\eta = 0.05$. In the case of RLS connectionist algorithm, the covariance matrices had the following initial values: $C^i(0) = 1000$; or $C^{ij}(0) = 1000$. In the case of gradient approximation the learning rate constant was $\alpha^3 = 10$. In Figure 2.16 the square criterion of learning in the case of trajectory tracking with standard BP and RLS is shown. Also, in Figure 2.17 a comparison of the BP method with the EKF method is given. In Figures 2.18 and 2.19, the position error for the first DOF is presented using BP, RLS and EKF algorithms. The figures show the errors in the first and 10th training trial. The above figures show that tracking errors exhibited considerable decrease with repetitive trials using RLS or EKF methods. The

Table 2.2. Parameters of D.C. actuators for robot MANUTEC r3

Actuator	1	2	3	4	5	6
Mechanical constant $[Nm/A]$	126.	252.	72.	24.8	21.4	8.6
Electrical constant $[Vs/rad]$	2.5	5.	1.428	2.2	1.74	2.027
M.inertia of rotor $[kgm^2]$	0.0013	0.0013	0.0013	0.00016	0.00018	0.00004
Viscous coefficient $[Nms/rad]$	1132.979	1091.	411.6	108.	112.6	3.6
Rotor resistance $[\Omega]$	1.	1.	1.	1.	1.	1.
Voltage saturation $[V]$	7.5	7.5	7.5	7.5	7.5	7.5
Reduction ratio	105.	210.	60.	99.	79.2	99.
Reducer efficiency	1.	1.	1.	1.	1.	1.
Rotor inductivity $[H]$	0.001	0.001	0.001	0.001	0.001	0.001

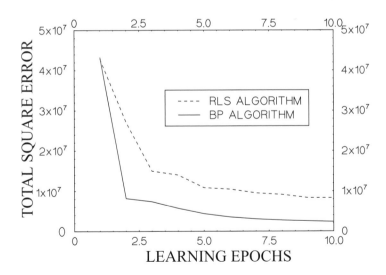

Figure 2.16. Comparison of BP and RLS algorithm

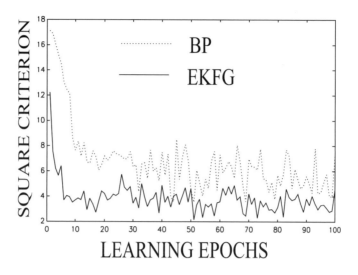

Figure 2.17. Comparison of BP and EKF algorithm

result also shows the better learning of the dynamic robot model and significant reduction of learning time in comparison with standard BP algorithm. Finally, in Figure 2.20 the position error for the first DOF in the case of trajectory tracking with learning and without learning is shown. It is evident that in the case of learning the robot trajectory tracking is much better than in the case without learning. The results presented in this section indicate the feasibility of using a highly efficient feedforward connectionist approach to learning complex input-output relations of robot dynamics. The important feature of new learning control structures is fast convergence properties, because the problem of adjusting the weights of internal hidden units in multilayer perceptrons is considered as a problem of estimating parameters by well-known identification methods. With these algorithms the learning rate is time dependent, which, in addition to a sufficiently convergent solution, ensures faster learning than the generalized delta rule in standard back propagation algorithms.

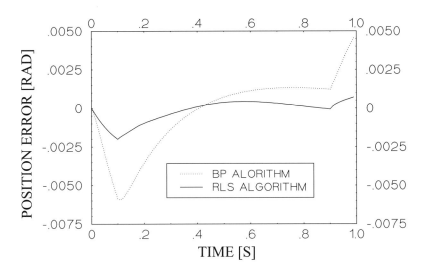

Figure 2.18. Position error - comparison of BP and RLS algorithms

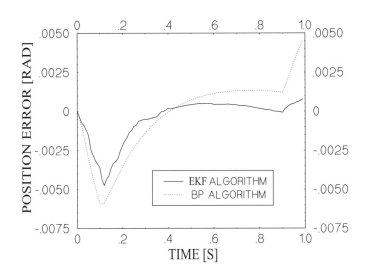

Figure 2.19. Position error - comparison of BP and EKF algorithms

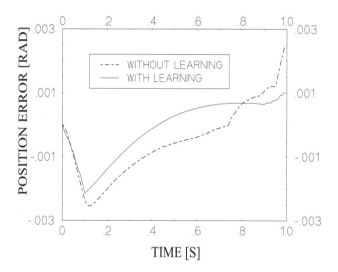

Figure 2.20.   Position error - with learning and without learning

Chapter 3

FUZZY LOGIC APPROACH IN ROBOTICS

1. Introduction

The basic principles of fuzzy control were formulated by Lotfi Zadeh in his papers from 1968, 1972 and 1973 [397], [399], [400]. The heart of this idea is to describe control strategy in linguistic terms. For instance, one possible control strategy of a single-input-single-output system can be described via a set of control rules:

if (error is positive and error change is positive) then
    control change = negative
else if (error is positive and error change is negative) then
      control change = zero
else if (error is negative and error change is positive) then
    control change = zero
else if (error is negative and error change is negative) then
    control change = positive

Further refining of the strategy might take into account the cases when, e.g., the error and error change are *small* or *big*. By such a procedure it could be possible to describe the control strategy used, e.g., by a trained operator when controlling a system manually.

Statements in natural language are intrinsically imprecise due to the imprecise manner of human reasoning. Development of techniques for modelling the imprecise statements is one of the main issues in the implementation of an automatic control system based on using linguistic control rules. With fuzzy controllers, modelling of linguistic control rules (as well as derivation of control action on the basis of the given set of rules and known state of the controlled system) is based on the theory of fuzzy sets, introduced by Zadeh in 1965 [403].

71

In 1974, Mamdani described the first application of the fuzzy set theory to automatic control [231]. However, almost ten years passed before the application of fuzzy logic in this area aroused a broader interest. At the beginning of the '90s, the number of reported fuzzy applications had an exponential increase (Figure 3.1). Current applications based on fuzzy control appear in such diverse areas as automatic control of trains, road cars, cranes, lifts, nuclear plants, home appliances, etc. Commercial applications in robotics still do not exist. However, numerous research efforts promise that fuzzy robot control systems can be expected soon, notably in the fields of robotized part processing, assembly, mobile robots, and robot vision systems.

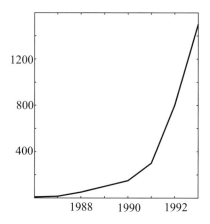

Figure 3.1.   Estimated number of commercial applications of fuzzy systems

Thanks to its ability to manipulate imprecise and incomplete data, fuzzy logic offers the possibility of incorporating expert experience into automatic control systems. Fuzzy logic has already proven itself useful in cases where the process is too complex to be analyzed by conventional quantitative techniques, or where the available information is qualitative, imprecise, or unreliable. Considering that it is based on precise mathematical theory, fuzzy logic additionally offers a possibility of integrating heuristic methods with conventional techniques for the analysis and synthesis of automatic control systems, thus facilitating further refinement of fuzzy control based systems.

2.   Mathematical foundations

2.1   Fuzzy sets

At the heart of the fuzzy set theory is the notion of fuzzy sets, which are used to model statements in natural (or artificial) language. Fuzzy set is a generalization of the classical (crisp) sets. The classical set concept assumes the possibility of dividing the particles of some universe into two partitions: particles that are members of the given set, and those which are not. This partition process can be described by means of the characteristic *membership function*. For a given universe of discourse X and the given set A, the membership function $\mu_A(\cdot)$ assigns a value to each particle $x \in X$ so that:

$$\mu_A(x) = \begin{cases} 1 & \text{if } x \in A \\ 0 & \text{otherwise} \end{cases} \tag{3.1}$$

With fuzzy sets, it is allowed that the boundary between the members and non-members of the set is not strict. This softening of the boundary is defined mathematically using the *membership degree function*, which assigns to each particle a value that indicates the degree of membership to the given set (see Figure 3.2). Accordingly, the fuzzy set $\tilde{A}$ in the universe of discourse X is defined by its degree of the membership function $\mu_{\tilde{A}}(\cdot)$ of the form:

$$\mu_{\tilde{A}} : X \mapsto [0, 1]. \tag{3.2}$$

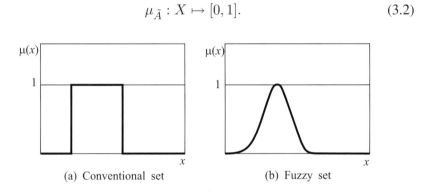

(a) Conventional set   (b) Fuzzy set

Figure 3.2.   Membership functions of conventional and fuzzy sets

For each fuzzy set, its support can be defined. The support of the fuzzy set $\tilde{A}$ is an ordinary set A which contains all the elements from the universe X with nonzero membership degrees in $\tilde{A}$:

$$\text{supp}(\tilde{A}) = \{x \in X : \mu_{\tilde{A}}(x) > 0\} \tag{3.3}$$

The notion of support allows one to formally define an empty fuzzy set: *empty fuzzy set* is a fuzzy set with empty support.

It is customary to represent fuzzy sets via fuzzy singletons: *fuzzy singleton* is such a fuzzy set for which its support is a single particle x from the universe X. If the fuzzy set $\tilde{A}$ has a finite support $\text{supp}(\tilde{A}) = \{x_1, x_2, \ldots, x_n\}$ with the degrees of membership $\mu_{\tilde{A}}(x_i)$, $i = 1, 2, \ldots, n$, such a fuzzy set is conveniently written as:

$$\tilde{A} = \mu_{\tilde{A}}(x_1)/x_1 + \mu_{\tilde{A}}(x_2)/x_2 + \cdots + \mu_{\tilde{A}}(x_n)/x_n = \sum_{i=1}^{n} \mu_{\tilde{A}}(x_i)/x_i \quad (3.4)$$

Here, the plus sign indicates that the pairs $\mu_{\tilde{A}}(x_i)/x_i$ collectively form the definition of the fuzzy set $\tilde{A}$. If the universe X is an interval of real numbers, then the following notation for the fuzzy set $\tilde{A}$ in X is customary:

$$\tilde{A} = \int_{X} \mu_{\tilde{A}}(x)/x \quad (3.5)$$

The notions of fuzzy subsets and equality between fuzzy sets are also defined in terms of membership degree functions. A fuzzy set $\tilde{A}$ is said to be a *subset* of $\tilde{B}$ if all particles $x \in X$ have degrees of membership to $\tilde{A}$ lower than or equal to their degrees of membership to $\tilde{B}$:

$$\tilde{A} \subseteq \tilde{B} \text{ iff } \mu_{\tilde{A}}(x) \leq \mu_{\tilde{B}}(x) \text{ for all } x \in X \quad (3.6)$$

Two fuzzy sets are *equal* if their membership functions are equal for all elements in the universe of discourse:

$$\tilde{A} = \tilde{B} \text{ iff } \mu_{\tilde{A}}(x) = \mu_{\tilde{B}}(x) \text{ for all } x \in X \quad (3.7)$$

An important class of fuzzy sets is the normalized fuzzy sets. A fuzzy set $\tilde{A}$ is said to be normalized if its height $h(\tilde{A})$, defined as the largest degree of membership that is attained by elements in its support, is equal to one:

$$h(\tilde{A}) \equiv \max_{X} \mu_{\tilde{A}}(x) = 1 \quad (3.8)$$

The value $m \in X$ for which $\mu_{\tilde{A}}(m) = h(\tilde{A})$ is called the *modal value* of the fuzzy set.

A fuzzy set $\tilde{A}$ in the Euclidean space R^n is *convex* if, for any vectors $\mathbf{x}, \mathbf{y} \in R^n$, the following is valid:

$$\mu_{\tilde{A}}(\lambda \mathbf{x} + (1 - \lambda)\mathbf{y}) \geq \min[\mu_{\tilde{A}}(\mathbf{x}), \mu_{\tilde{A}}(\mathbf{y})] \quad (3.9)$$

The fuzzy sets that are normalized, convex and, additionally, have a piecewise continuous membership degree function, are denoted as *fuzzy intervals*. A special class of fuzzy intervals are *fuzzy numbers:* a fuzzy number is a fuzzy interval with unique modal value. The concept of fuzzy numbers makes a basis of fuzzy arithmetic, which may be considered as a generalization of classical arithmetic. Examples of membership functions of normalized, convex fuzzy sets, and fuzzy numbers are shown in Figure 3.3.

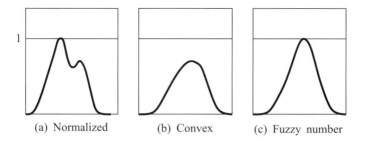

(a) Normalized   (b) Convex   (c) Fuzzy number

Figure 3.3. Examples of fuzzy sets

2.2   Operations on fuzzy sets

The basic principle of the generalization of classical mathematical concepts for the field of fuzzy sets is known as the principle of extension [401]. Formally, given a function $f : X \mapsto Y$, mapping elements of the ordinal set X into elements of the set Y, and an arbitrary fuzzy set $\tilde{A} \in \tilde{P}(X)$, e.g.:

$$\tilde{A} = \mu_1/x_1 + \mu_2/x_2 + \cdots + \mu_n/x_n \tag{3.10}$$

the principle of extension states that the following relation has to be preserved:

$$\begin{aligned} f(\tilde{A}) &\overset{\triangle}{=} f(\mu_1/x_1 + \mu_2/x_2 + \cdots + \mu_n/x_n) \\ &= \mu_1/f(x_1) + \mu_2/f(x_2) + \cdots + \mu_n/f(x_n) \end{aligned} \tag{3.11}$$

In other words, operations on fuzzy sets should preserve the important properties of operations on classical sets. Unfortunately, it turns out that it is not possible to make such a definition of basic fuzzy set operations that would preserve all important properties of the corresponding operations on classical sets. For example, it is shown that arbitrary fuzzy complement, union, and intersection operations satisfying the law of contradiction and the law of excluded middle, are not distributive. Therefore, a choice of basic fuzzy set operations has to be made by considering the context in which these operations shall be carried out. The most often used set of basic *standard operations* of fuzzy set theory is (see Figure 3.4):

$$
\begin{aligned}
\textit{Complement:} \quad & \mu_{\overline{\tilde{A}}}(x) &=& \quad 1 - \mu_{\tilde{A}}(x) \\
\textit{Union:} \quad & \mu_{\tilde{A}\cup\tilde{B}}(x) &=& \quad \max[\mu_{\tilde{A}}(x), \mu_{\tilde{B}}(x)] \\
\textit{Intersection:} \quad & \mu_{\tilde{A}\cap\tilde{B}}(x) &=& \quad \min[\mu_{\tilde{A}}(x), \mu_{\tilde{B}}(x)]
\end{aligned}
$$

The fuzzy set theory that is based on the operators defined this way is usually referred to as *possibility theory*. However, in some situations, different

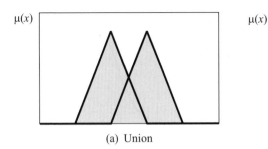
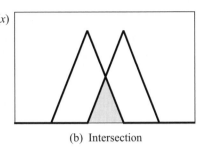

(a) Union   (b) Intersection

Figure 3.4.   Standard operations on fuzzy sets

definitions of basic fuzzy set operators are preferable. For example, a union $\tilde{A} \cup \tilde{B}$ intuitively represents a disjunction of the concepts represented by $\tilde{A}$ and $\tilde{B}$. Additionally, the notion of union normally implies a certain level of interchangeability between the concepts represented by its arguments. On the other hand, the standard union max operator is rigid in the sense that it does not assume such an interchangeability. If the union were specified by the function:

$$f_u : [0,1] \times [0,1] \mapsto [0,1] \tag{3.12}$$

that assigns a value $f_u[\mu_{\tilde{A}}(x), \mu_{\tilde{B}}(x)]$ to a given pair of membership degrees $\mu_{\tilde{A}}(x)$ and $\mu_{\tilde{B}}(x)$, then the intuitive meaning of the union would imply the following relation:

$$f_u[\mu_{\tilde{A}}(x), \mu_{\tilde{B}}(x)] \geq \max[\mu_{\tilde{A}}(x), \mu_{\tilde{B}}(x)] \tag{3.13}$$

It is evident that standard union operation, defined as $\max[\mu_{\tilde{A}}(x), \mu_{\tilde{B}}(x)]$, yields the lowest possible degree of membership. For this reason, in some cases alternative formulations are used in place of the max operator. All potential formulations $f_u(\cdot)$ are required to satisfy the minimum axiomatic conditions:

U1. *Boundary conditions:* $f_u(0,0) = 0$ and $f_u(0,1) = f_u(1,0) = f_u(1,1) = 1$;

U2. *Commutativity:* $f_u(x,y) = f_u(y,x)$;

U3. *Monotonicity:* if $x \leq x'$ and $y \leq y'$, then $f_u(x,y) \leq f_u(x',y')$;

U4. *Associativity:* $f_u(f_u(x,y),z) = f_u(x,f_u(y,z))$.

The functions satisfying these axioms are called *triangular conorms (t-conorms)*. Evidently, the standard union operation is a t-conorm. Other t-conorms are proposed as well, such as algebraic sum, bounded sum, etc.The fuzzy intersection $\tilde{A} \cap \tilde{B}$ intuitively denotes a conjunction of the concepts represented by $\tilde{A}$ and

$\tilde{B}$. As in the case of union, the intersection operation can be specified using the function:

$$f_i : [0, 1] \times [0, 1] \mapsto [0, 1] \qquad (3.14)$$

The minimum axiomatic skeleton that the functions $f_i(\cdot)$ have to satisfy in order to qualify as candidates for the definition of fuzzy intersection, consists of the conditions:

I1. *Boundary conditions:* $f_i(1, 1) = 1$ and $f_i(0, 0) = f_i(0, 1) = f_i(1, 0) = 0$;

I2. *Commutativity:* $f_i(x, y) = f_i(y, x)$;

I3. *Monotonicity:* if $x \le x'$ and $y \le y'$, then $f_i(x, y) \le f_i(x', y')$;

I4. *Associativity:* $f_i(f_i(x, y), z) = f_i(x, f_i(y, z))$.

The functions satisfying the axioms I1-I3 are called *triangular norms (t-norms)*. Obviously, the standard min operation is a t-norm.

Analogously to the case of the union, the intersection of fuzzy sets normally implies a certain level of requirement for simultaneous satisfaction of the concepts represented by its arguments. On the other hand, the standard min operation is rigid in the sense that it does not account for the benefits of simultaneous memberships. Hence, alternative t-norms are proposed, in which different intensities of intersections are achieved: algebraic product, bounded product, etc. Standard min operation represents the upper bound of the possible intersection operations (the weakest intersection).

2.3 Fuzzy relations

Fuzzy relation is a generalization of the classical concept of relation between elements of two or more sets. Additionally, fuzzy relations allow one to specify different levels of strength of association between individual elements. The levels of association are represented via degrees of membership to the fuzzy relation, in the same manner as the degree of membership to a fuzzy set is represented.

Formally, a fuzzy relation between elements of the ordinary sets $X_1, X_2, ..., X_n$ is a fuzzy subset $\tilde{R} = \tilde{R}(X_1, X_2, .., X_n)$ of the Cartesian product $X_1 \times X_2 \times .. \times X_n$ and it is defined by the membership degree function:

$$\mu_{\tilde{R}} : X_1 \times X_2 \times ... \times X_n \mapsto [0, 1] \qquad (3.15)$$

Thus, the tuples $\mathbf{x} = (x_1, x_2, .., x_n) \in X_1 \times X_2 \times .. \times X_n$ may have different degrees of membership $\mu_{\tilde{R}}(x_1, x_2, .., x_n) \in [0, 1]$ to the fuzzy relation.

When the sets $X_1, X_2, .., X_n$ are finite, the fuzzy relation $\tilde{R}(X_1, X_2, .., X_n)$ is conveniently represented by the n-dimensional *membership matrix,* whose elements show the degree to which the individual tuples belong to a given fuzzy relation. For instance, the binary fuzzy relation $\tilde{R}(X, Y)$ between the sets $X = \{x_1, \ldots, x_n\}$ and $Y = \{y_1, \ldots, y_m\}$ is conveniently represented by the matrix:

$$\tilde{\mathbf{R}} = \begin{bmatrix} \mu_{x_1,y_1} & \cdots & \mu_{x_1,y_m} \\ \vdots & \ddots & \vdots \\ \mu_{x_n,y_1} & \cdots & \mu_{x_n,y_m} \end{bmatrix} \tag{3.16}$$

For the given family of sets $\tilde{A}_1, \tilde{A}_2, \ldots, \tilde{A}_n$, defined in the universes $X_1 \times X_2 \times \cdots \times X_n$, the Cartesian product of fuzzy sets:

$$\tilde{A}_1 \times \tilde{A}_2 \times \cdots \times \tilde{A}_n \tag{3.17}$$

is a fuzzy set in the universe of discourse $X_1 \times X_2 \times \cdots \times X_n$. Consequently, the Cartesian product is an n-ary fuzzy relation with the degree of membership function defined by:

$$\mu_{\tilde{A}_1 \times \tilde{A}_2 \times \cdots \times \tilde{A}_n}(x_1, x_2, \ldots, x_n) = \mu_{\tilde{A}_1}(x_1) \star \mu_{\tilde{A}_2}(x_2) \star \ldots \star \mu_{\tilde{A}_n}(x_n) \tag{3.18}$$

for all $x_1 \in X_1, x_2 \in X_2, \ldots, x_n \in X_n$, where the sign $\star$ denotes one of triangular norms (i.e., the intersection operation).

Among the operations over fuzzy relations, compositions of binary relations are of special significance. For the ordinary binary relations $P(X, Y)$ and $Q(Y, Z)$, defined in the common set Y, the *composition* of P and Q:

$$R(X, Z) = P(X, Y) \circ Q(Y, Z) \tag{3.19}$$

is defined as a subset $R \subseteq X \times Z$ such that:

$$(x, z) \in R \text{ iff there exists } y \in Y \text{ for which } (x, y) \in P \text{ and } (x, z) \in Q \tag{3.20}$$

The concept of composition is extended to fuzzy relations in a number of ways aimed at preserving important properties of the corresponding compositions of classical relations. The most important types of compositions of binary fuzzy relations are:

■ *max-min composition.* Denoted by $\tilde{P}(X, Y) \circ \tilde{Q}(Y, Z)$, this operation is defined by:

$$\mu_{\tilde{P} \circ \tilde{Q}}(x, z) = \max_{y \in Y} \min[\mu_{\tilde{P}}(x, y), \mu_{\tilde{Q}}(y, z)] \tag{3.21}$$

Thus, the strength of the relation between the elements x and z is equal to the strength of the strongest chain between these elements, whereas the strength of each chain x-y-z is equal to the strength of its weakest link.

- *max-product composition.* The composition is denoted by $\tilde{P}(X,Y) \odot \tilde{Q}(Y,Z)$ and defined via:

$$\mu_{\tilde{P} \odot \tilde{Q}}(x,z) = \max_{y \in Y}[\mu_{\tilde{P}}(x,y) \cdot \mu_{\tilde{Q}}(y,z)] \qquad (3.22)$$

The max-min and max-product compositions may be regarded as specialization of the more general *sup-star composition,* denoted by $\tilde{P}(X,Y) \circ \tilde{Q}(Y,Z)$ and defined via:

$$\mu_{\tilde{P} \circ \tilde{Q}}(x,z) = \sup_{y \in Y}[\mu_{\tilde{P}}(x,y) \star \mu_{\tilde{Q}}(y,z)] \qquad (3.23)$$

where the sign $\star$ represents any triangular norm, and the sup operator denotes supreme (the lowest upper bound).

Compositions of binary relations in finite sets may be efficiently realized using membership matrices. For example, the composition $\tilde{P} \circ \tilde{Q}$ can be calculated as a matrix product:

$$\tilde{\mathbf{P}} \cdot \tilde{\mathbf{Q}} \qquad (3.24)$$

where multiplication is replaced by the min, and addition by the max operator.

2.4   Fuzzy logic

Fuzzy logic is a discipline engaged in formal principles of approximate reasoning [402]. Its main issue is the modelling of imprecise modes of human reasoning under conditions characterized by unreliability and imprecision, whereby the theory of fuzzy sets is used as a basic methodology.

Fuzzy logic is an extension to classical logic in which the basic objects are logical propositions that may take one of the possible values of truth: true or false, i.e., 1 or 0. Contrary to classical formal systems, fuzzy logic allows one to evaluate the truth of a proposition as e.g., a real number in the interval $[0, 1]$. The basis of fuzzy logic is the theory of fuzzy sets. For example, the characterization of the fuzzy set $\tilde{A}$ with the membership function $\mu_{\tilde{A}}(x)$, $x \in X$, can be interpreted as the truth value of the proposition:

$$x \text{ is an element of } \tilde{A} \qquad (3.25)$$

Aimed at enabling work with imprecise propositions, fuzzy logic permits the use of:

- *Fuzzy predicates:* truth values of an imprecise predicate $P(x)$ (e.g., x is *small, big,* etc.) can be described for any $x \in X$ via the fuzzy set with the membership function $\mu_P(x)$, determined by the predicate $P(\cdot)$;

- *Fuzzy truth values:* fuzzy sets, defined on the interval $[0, 1]$, can be used to describe different levels of truth (e.g., *fairly true, completely false*, etc.);

- *Fuzzy quantifiers:* besides the usual quantifiers from classical logic ($\forall, \exists$), imprecise statements may contain imprecise quantifiers (e.g., *sometimes, almost always)* that can be represented by fuzzy numbers;

- *Fuzzy modifiers:* different forms of fuzzy modifiers (*probably, fairly*, etc.) can be described by utilizing special operations on fuzzy sets, representing the modified propositions.

The central problem of the *quantitative fuzzy semantics* is calculating the meaning of *linguistic variables*, i.e., the variables whose values are sentences in a specific (natural or artificial) language [398]. The linguistic variable can be regarded as a variable whose value is a fuzzy number (the *meaning* of the variable) or as a variable whose values are linguistically defined (the *label* of the variable) [401]. Generally, the label of linguistic variable is obtained by concatenating the terms of the language according to some rules. In simple cases, these terms can be divided into four categories:

1 *Primary terms* that represent labels of specific fuzzy sets;

2 Negation *not* and connectives *or* and *and;*

3 *Modifiers* that modify the basic concept to which they are applied (e.g. *very, extremely,* etc.);

4 *Markers* such as parentheses.

Negation *not* and connectives *or* and *and* may be considered as labels of the corresponding operations on fuzzy sets:

- *Complement* $\overline{\tilde{A}}$, which represents the fuzzy concept "*not $\tilde{A}$*";

- *Union* $\tilde{A} \cup \tilde{B}$, which represents the fuzzy concept "$\tilde{A}$ *or* $\tilde{B}$";

- *Intersection* $\tilde{A} \cap \tilde{B}$, which represents the fuzzy concept "$\tilde{A}$ *and* $\tilde{B}$".

Linguistic modifiers can be expressed via specific operations on the fuzzy set $\tilde{A}$ that describes the basic concept, e.g.:

- *Exponent* $\tilde{A}^{\alpha}$, defined via:

$$\mu_{\tilde{A}^{\alpha}}(x) = [\mu_{\tilde{A}}(x)]^{\alpha} \tag{3.26}$$

- *Concentration,* defined via:

$$\mathrm{con}(\tilde{A}) = \tilde{A}^2 \tag{3.27}$$

The operation of concentration may be interpreted as "*very $\tilde{A}$*" and its effect is a large reduction of degrees of membership of those values of x that already have a small degree of membership $\mu_{\tilde{A}}(x)$ to the basic concepts $\tilde{A}$, with additional small reduction for those x with high membership $\mu_{\tilde{A}}(x)$.

- *Dilatation,* defined via:

$$\text{dil}(\tilde{A}) = \tilde{A}^{0.5} \tag{3.28}$$

The operation of dilatation can be described as "*more or less $\tilde{A}$*" and its effect is opposite to that of concentration.

- *Contrast intensification* int($\tilde{A}$), defined via:

$$\mu_{\text{int}(\tilde{A})}(x) = \begin{cases} 2[\mu_{\tilde{A}}(x)]^2 & \text{for } \mu_{\tilde{A}}(x) \leq 0.5 \\ 1 - 2[1 - \mu_{\tilde{A}}(x)]^2 & \text{for } \mu_{\tilde{A}}(x) > 0.5 \end{cases} \tag{3.29}$$

This operation has the consequence of increasing the values $\mu_{\tilde{A}}(x)$ that are above the *crossover point* 0.5 and reducing the values that are below this point.

Examples of linguistic modifiers are illustrated in Figure 3.5.

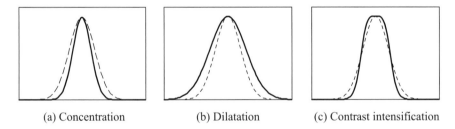

(a) Concentration   (b) Dilatation   (c) Contrast intensification

Figure 3.5. Examples of linguistic modifiers

Among various forms of fuzzy propositions, fuzzy implications are of special importance. *Fuzzy implication* is a statement of the form:

$$\tilde{A}(x) \Rightarrow \tilde{B}(y) \tag{3.30}$$

or, equivalently,

$$\text{if } \tilde{A}(x) \text{ then } \tilde{B}(y) \tag{3.31}$$

where $x \in X, y \in Y$ are linguistic variables and $\tilde{A}(\cdot), \tilde{B}(\cdot)$ are fuzzy predicates in the universes of discourse X and Y, respectively. Essentially, such a statement describes the fuzzy relation:

$$\tilde{R}_{\Rightarrow}(X, Y) \subseteq X \times Y \tag{3.32}$$

between the two fuzzy sets, i.e., between the equivalent fuzzy propositions $\tilde{A}(x)$ and $\tilde{B}(y)$.

Fuzzy implication is important because of its role in automatic inferencing. The two basic *fuzzy inference rules* based on fuzzy implication, are:

- *Generalized modus ponens:* $(\tilde{A}'(x) \wedge (\tilde{A}(x) \Rightarrow \tilde{B}(y))) \Rightarrow \tilde{B}'(y)$;

- *Generalized modus tollens:* $(\tilde{B}'(y) \wedge (\tilde{A}(x) \Rightarrow \tilde{B}(y))) \Rightarrow \tilde{A}'(x)$;

The generalized modus ponens is closely related to the mechanism of forward inferencing (data-driven inference) and it reduces to the classical modus ponens when $\tilde{A}' = \tilde{A}$ and $\tilde{B}' = \tilde{B}$. Analogously, the generalized modus tollens is closely related to the mechanism of backward inferencing (goal-driven inference) and it reduces to the classical modus tollens when $\tilde{A}' = \overline{\tilde{A}}$ and $\tilde{B}' = \overline{\tilde{B}}$.

A basic technique that lies at the heart of most implementation of automatic fuzzy inference is the *compositional rule of inference,* proposed by Zadeh [400]. According to this rule, the binary fuzzy relation $\tilde{R}$ from X to Y and the fuzzy set $\tilde{x} \subseteq X$ induce the fuzzy set $\tilde{y} \subseteq Y$, determined by the sup-star composition

$$\tilde{y} = \tilde{x} \circ R \qquad (3.33)$$

in which $\tilde{x}$ plays the role of a unary fuzzy relation. When the setting $\tilde{R} = \tilde{A} \Rightarrow \tilde{B}$, $\tilde{x} = \tilde{A}'$, and $\tilde{y} = \tilde{B}'$ in the compositional rule, the rule becomes an implementation of the generalized modus ponens:

$$\tilde{B}' = \tilde{A}' \circ (\tilde{A} \Rightarrow \tilde{B}). \qquad (3.34)$$

If $\tilde{A}, \tilde{A}', \tilde{B}, \tilde{B}'$ are nonfuzzy and $\tilde{A}' = \tilde{A}$, the compositional rule of inference becomes:

$$\tilde{B}' = \tilde{A} \circ (\tilde{A} \Rightarrow \tilde{B}) = \tilde{B}. \qquad (3.35)$$

Thus, the compositional rule can be regarded as an approximate extension, i.e., a fuzzy generalization of modus ponens: the more different $\tilde{A}'$ is from $\tilde{A}$, the less sharply defined is $\tilde{B}'$.

Bearing in mind the significance of fuzzy implication, a number of distinct fuzzy implication functions have been proposed for its implementation. The proposed functions can be divided into five families [69]:

1 Material implication: $\tilde{A} \Rightarrow \tilde{B} \overset{\triangle}{=} \overline{\tilde{A}} \cup \tilde{B}$;

2 Implication in propositional calculus: $\tilde{A} \Rightarrow \tilde{B} \overset{\triangle}{=} \overline{\tilde{A}} \cup (\tilde{A} \cap \tilde{B})$;

3 Extended implication in propositional calculus: $\tilde{A} \Rightarrow \tilde{B} \overset{\triangle}{=} (\overline{\tilde{A}} \cap \overline{\tilde{B}}) \cup \tilde{B}$;

4 Generalization of modus ponens: $\tilde{A} \Rightarrow \tilde{B} \overset{\triangle}{=} \sup_{\tilde{C}} \{\tilde{C} : \tilde{C} \cap \tilde{A} \subseteq \tilde{B}\}$;

5 Generalization of modus tollens: $\tilde{A} \Rightarrow \tilde{B} \stackrel{\triangle}{=} \inf_{\tilde{C}} \{\tilde{C} : \tilde{C} \cup \tilde{B} \subseteq \tilde{A}\}$.

Several authors have analyzed the axiomatic requirements and criteria for the selection of appropriate functions for the implementation of fuzzy implication [22], [85], [204]. One of the widely accepted definitions is the *standard fuzzy implication,* representing an implementation of generalized modus ponens, in which the standard union and intersection operators are used:

$$\mu_{\tilde{A} \Rightarrow \tilde{B}}(x, y) = \begin{cases} 1 & \text{for } \mu_{\tilde{A}}(x) \leq \mu_{\tilde{B}}(y) \\ \mu_{\tilde{B}}(y) & \text{for } \mu_{\tilde{A}}(x) > \mu_{\tilde{B}}(y) \end{cases} \tag{3.36}$$

3.   Fuzzy controller

Fuzzy control approaches the control problem in a radically different way compared to the traditional model-based techniques. Instead of precise mathematical models, fuzzy control uses an imprecise and incomplete description of the process and/or the manner in which the system is controlled by human operators, whereby the theory of fuzzy sets is used as a principle tool.

Fuzzy controller consists of four basic components (see Figure 3.6): condition (fuzzification) interface, knowledge base, inference mechanism, and action interface.

The block denoted as condition interface performs measurement of the input (state) variables:

$$\mathbf{x} = [x_1, \ x_2, \ \ldots, \ x_n]^T \tag{3.37}$$

of the controlled process and translates them into the fuzzy linguistic terms $X_1, X_2, \ldots, X_n$ that are represented by the fuzzy sets $\tilde{x}_1, \tilde{x}_2, \ldots, \tilde{x}_n$ in the appropriate universes of discourse $U_1, U_2, \ldots, U_n$, respectively. The obtained fuzzy values constitute the *fuzzy state* of the process:

$$\tilde{\mathbf{x}} = [\tilde{x}_1, \ \tilde{x}_2, \ \ldots, \ \tilde{x}_n]^T \tag{3.38}$$

in the state space $\mathbf{U} = U_1 \times U_2 \times \cdots \times U_n$. The fuzzy state variables are further used in the evaluation of fuzzy control rules.

The knowledge base consists of control rules and a fuzzy set definition base. The definition base provides the definitions necessary to characterize fuzzy control rules and manipulation of fuzzy data. The rule base consists of heuristic fuzzy control rules that describe the control goals and control policy. A *fuzzy control rule* is a fuzzy conditional statement (fuzzy implication) in which the antecedent is a condition and the consequence is a control action. Thus, the

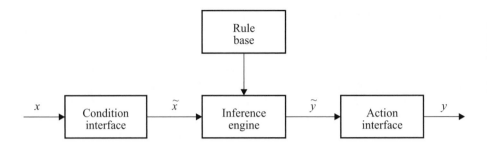

Figure 3.6.   Components of fuzzy controller

rule base can be represented as:

R_1 :   if

X_1 is A_{11} and ... and X_n is A_{1n}

then

$Y_1 = B_{11}$ and ... and $Y_m = B_{1m}$

R_2 :   if

X_1 is A_{21} and ... and X_n is A_{2n}

then

$Y_1 = B_{21}$ and ... and $Y_m = B_{2m}$

$\vdots$

R_r :   if

X_1 is A_{r1} and ... and X_n is A_{rn}

then

$Y_1 = B_{r1}$ and ... and $Y_m = B_{rm}$

where:

$$
\begin{aligned}
X_1, X_2, \ldots, X_n \;\; &= \;\; \text{Labels of fuzzy state variables } \tilde{x}_1, \tilde{x}_2, \ldots, \tilde{x}_n \\
&\quad\; \text{in the universes } U_1, U_2, \ldots, U_n; \\
Y_1, Y_2, \ldots, Y_m \;\; &= \;\; \text{Labels of fuzzy actions } \tilde{y}_1, \tilde{y}_2, \ldots \tilde{y}_n \\
&\quad\; \text{in the universes } V_1, V_2, \ldots, V_m; \\
A_{ki} \text{ and } B_{kj} \;\; &= \;\; \text{Labels of fixed linguistic values represented by} \\
&\quad\; \text{the fuzzy sets } \tilde{a}_{ki} \subseteq U_i, \tilde{b}_{kj} \subseteq V_j.
\end{aligned}
$$

The rules R_k, $k = 1, 2, \ldots, r$ are mutually interconnected via the implicit connectives *also*. Each control rule is implemented by the fuzzy relation $\tilde{\mathbf{R}}_k$ in $\mathbf{U} \times \mathbf{V}$, where $\mathbf{U} = U_1 \times U_2 \times \cdots \times U_m$ and $\mathbf{V} = V_1 \times V_2 \times \cdots \times V_m$. The rule base is an aggregate of individual rules. By integration of the particular relations $\tilde{\mathbf{R}}_k$, $k = 1, 2, \ldots, r$, the aggregate relation of the whole rule base is obtained as:

$$\tilde{\mathbf{R}} \subseteq \mathbf{U} \times \mathbf{V} \tag{3.39}$$

The block designated as inference mechanism is responsible for the evaluation of control rules. Evaluation is commonly carried out using the sup-star compositional rule [400]:

$$\tilde{\mathbf{y}} = \tilde{\mathbf{x}} \circ \tilde{\mathbf{R}} \qquad (3.40)$$

The result is the *fuzzy control action* $\tilde{\mathbf{y}} = [\tilde{y}_1 \; \tilde{y}_2 \; \ldots \; \tilde{y}_m]^T$ in the universe $\mathbf{V}$ of possible control actions. Within the action interface, the fuzzy action $\tilde{\mathbf{y}}$ is converted into the defuzzified action $\mathbf{y} = (y_1, y_2, \ldots, y_m)$.

3.1 Condition interface

The task of the condition interface is (1) to perform *scale mapping* that transfers the range of values of input variables into the corresponding universes of discourse and (2) to perform *fuzzification* that converts crisp inputs into fuzzy sets.

The most frequent fuzzification strategy consists of transforming the measured value x into a fuzzy singleton $\tilde{x}$. Thus, the input x is interpreted as a fuzzy set $\tilde{x}$ with the membership function equal to zero at all points $u \in U$, except for the point u_0, where $\mu_{\tilde{x}}(u_0) = 1$.

3.2 Fuzzy set definition base

The fuzzy set definition base contains definitions of the fuzzy sets $\tilde{a}_{ki}$ and $\tilde{b}_{kj}$ ($i = 1, 2, \ldots, n$, $j = 1, 2, \ldots, m$, $k = 1, 2, \ldots, r$) that correspond to the linguistic labels A_{ki} and B_{kj} appearing in the control rules. These fuzzy sets are frequently designated as *primary fuzzy sets*.

The universes of discourse for input and output control signals can be discrete or continuous. In order to attain a more efficient manipulation with fuzzy sets, two basic transformations are commonly applied to the input/output spaces:

- *Normalization,* by which the universe of discourse U is transformed into the normalized closed interval $U_N = [-1, +1]$. The transformation function $f_N(\cdot)$ may be linear or nonlinear and its synthesis assumes a priori knowledge on the possible range $U = [u_{\min}, u_{\max}]$ of the signal. For the case of the linear mapping,

$$f_N(u) = [(u - u_{\max}) + (u - u_{\min})]/(u_{\max} - u_{\min}) \qquad (3.41)$$

  by choosing appropriate nonlinear transformation, a uniform distribution of symmetric and mutually equal primary sets may be achieved.

- *Discretization (quantization),* by which the continuous universe U or U_N is partitioned into a finite number of segments:

$$\overline{u}_1 = [\hat{u}_0, \hat{u}_1], \overline{u}_2 = (\hat{u}_1, \hat{u}_2], \ldots, \overline{u}_q = (\hat{u}_{q-1}, \hat{u}_q], \qquad (3.42)$$

  specified by the quantization levels $u_{\min} = \hat{u}_0 < \hat{u}_1 < \cdots < \hat{u}_q = u_{\max}$. Each segment $\overline{u}_i, i = 1, 2, \ldots, q$ is treated as a generic element representing all the elements $u \in \overline{u}_i$. In this manner, fuzzy sets can now be defined by assigning degrees of membership to each generic element of the universe $\overline{U} = \{\overline{u}_1, \overline{u}_2, \ldots, \overline{u}_q\}$.

Quantization may also be linear or nonlinear. The number of quantization levels should be sufficiently large to ensure adequate approximation and yet be small to save memory space. In the majority of applications, the number of quantization levels is 16 to 32.

Primary fuzzy sets are usually represented by linguistic labels such as: NB — negative big, NM – negative medium, NS – negative small, ZE – zero, PS – positive small, PM – positive medium, and PB – positive big. The set of different labels:

$$A_i = \bigcup_{k=1}^{r} \{A_{ki}\} \tag{3.43}$$

is called *fuzzy input space* of the i-th input variable, $i = 1, 2, \ldots, n$. Analogously, the set of different labels:

$$B_j = \bigcup_{k=1}^{r} \{B_{kj}\} \tag{3.44}$$

is called *fuzzy control space* of the j-th control variable, $j = 1, 2, \ldots, m$. The number of different labels in fuzzy input space determines the number of possible control rules. Finding the optimal fuzzy partition of the input space is a difficult task and is usually performed in a heuristic way.

Depending on whether the underlying universe of discourse is continuous or discrete, primary fuzzy sets are specified using functional or numerical definition. For the case of a continuous universe, the commonly applied functional forms of membership functions are:

- *Triangular functions:* $\mu_f(x) = \max(1 - |1 + \frac{x - x_f}{\sigma_f}|, 0)$

- *Bell-shaped functions:* $\mu_f(x) = e^{-\frac{1}{2}\left(\frac{x - x_f}{\sigma_f}\right)^2}$

An example of triangular primary fuzzy sets is given in Figure 3.7. If the universe of the discourse is discrete, a fuzzy set is represented as a vector whose elements are values of the membership degree.

3.3   Control rules

A rule R_k in the rule base typically takes the form of a *state evaluation fuzzy control rule*:

$$R_k : \text{ if } (X_1 \text{ is } A_{k1} \text{ and } \ldots \text{ and } X_n \text{ is } A_{kn}) \text{ then } Y = B_k$$

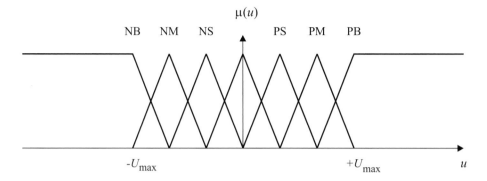

Figure 3.7. Primary fuzzy sets

Linguistic variables appearing on the left side of the implication are typically the process state, error (i.e., deviation from the desired state), and error change (i.e., time derivative of the error). The variable on the right side is usually the control output or change of the control output.

A more general form is *functional control rule*, where the premises and consequents are specified as (logical) functions:

$$R_k : \text{if } f_k(X_1 \text{ is } A_{k1}, \ldots, X_n \text{ is } A_{kn}) \text{ then } Y = g_k(X_1, X_2, \ldots X_n)$$

A popular form of functional rules are *Sugeno-type rules* [322], where the antecedent propositions are connected by fuzzy conjunction, and consequently, is a linear function of the input variables:

$$R_k \quad : \quad \text{if } X_1 \text{ is } A_{k1} \text{ and } \ldots \text{ and } X_n \text{ is } A_{kn} \text{ then}$$
$$y_k = g_{k0} + g_{k1}x_1 + \cdots + g_{kn}x_n$$

The main questions with the implementation of fuzzy controllers are related to the derivation and validation of control rules. Keeping in mind that fuzzy control is primarily efficient in cases when only qualitative and incomplete information is available, the rules are often derived in a heuristic way. Expert knowledge and imitation of the procedures employed by trained operators are commonly used. Adjustments of control parameters, aimed at improvements in the system performance, are often made using *ad hoc* procedures, which usually reduce to trial-and-error.

More recently, intensive investigations have been conducted on the development of systematic methods of deriving fuzzy control rules. Most of them use the notion of the *fuzzy process model*, i.e., the linguistic description of dynamic characteristics of the controlled process [331], [386]. The fuzzy model allows to approach identification of the linguistic structure and model parameters in a systematic manner. Based on the known fuzzy model, control rules can be generated for attaining optimal behaviour of the system.

An alternative solution lies in adding the fuzzy controller learning capabilities, i.e., the facilities to adapt general control rules to the actual situation. In principle, a fuzzy controller with learning capabilities has a hierarchical structure which consists of two rule bases: a general rule base and a base of meta-rules. The meta-rules exhibit human-like learning ability to create and modify the general rules based on the observed and desired performance characteristics of the system. The first such system with learning capabilities, a *self-organizing controller,* has been described by Procyk and Mamdani [279].

3.4   Inference mechanism

Consider the rule base:

$$
\begin{array}{ll}
R_1 : & \text{if } X_1 \text{ is } A_{11} \text{ and } \ldots \text{ and } X_n \text{ is } A_{1n} \text{ then } Y = B_1 \\
R_2 : & \text{if } X_1 \text{ is } A_{21} \text{ and } \ldots \text{ and } X_n \text{ is } A_{2n} \text{ then } Y = B_2 \\
\vdots & \\
R_r : & \text{if } X_1 \text{ is } A_{r1} \text{ and } \ldots \text{ and } X_n \text{ is } A_{rn} \text{ then } Y = B_r
\end{array}
$$

The antecedent of each rule $R_k, k = 1, 2, \ldots, r$ is interpreted as a fuzzy set:

$$
\tilde{\mathbf{a}}_k = \tilde{a}_{k1} \times \tilde{a}_{k2} \times \cdots \times \tilde{a}_{kn} \tag{3.45}
$$

in the product space $\mathbf{U} = U_1 \times U_2 \times \cdots \times U_n$, with the membership function $\mu_{\tilde{\mathbf{a}}_k}(\cdot)$ given for all $\mathbf{u} \in \mathbf{U}$ by:

$$
\mu_{\tilde{\mathbf{a}}_k}(\mathbf{u}) = \mu_{\tilde{\mathbf{a}}_k}(u_1, u_2, \ldots, u_n) = f_i(\mu_{\tilde{a}_{k1}}(u_1), \mu_{\tilde{a}_{k2}}(u_2), \ldots, \mu_{\tilde{a}_{kn}}(u_n)) \tag{3.46}
$$

where $f_i(\cdot)$ denotes any t-norm (intersection) function, such as min or algebraic product. Thus, the rule base can be represented in the form:

$$
\begin{array}{ll}
R_1 : & \text{if } \tilde{\mathbf{a}}_1 \text{ then } \tilde{b}_1 \\
R_2 : & \text{if } \tilde{\mathbf{a}}_2 \text{ then } \tilde{b}_2 \\
\vdots & \\
R_r : & \text{if } \tilde{\mathbf{a}}_r \text{ then } \tilde{b}_r
\end{array}
$$

where the antecedents are fuzzy sets in the universe $\mathbf{U}$, and the consequence are fuzzy sets in the universe V.

If the rule base is *complete* (i.e., it contains all possible fuzzy conditions and, additionally, for every input $\mathbf{u}$ there exists a *dominant rule* R_k with the applicability degree $\mu_{\tilde{\mathbf{a}}_k}(\mathbf{u})$ higher than some level, say 0.5), then such a base can be interpreted as a sequence of *fuzzy conditional statements:*

$$\text{if } \tilde{a}_1 \text{ then } \tilde{b}_1$$
$$\text{else if } \tilde{a}_2 \text{ then } \tilde{b}_2$$
$$\vdots$$
$$\text{else if } \tilde{a}_r \text{ then } \tilde{b}_r$$

It is natural to interpret the fuzzy conditional statements as Cartesian products, and connectives between the conditional statements as union. Thus, the relation represented by the rule base is naturally implemented as:

$$\tilde{\mathbf{R}} = \bigcup_{k=1}^{r} \tilde{\mathbf{R}}_k = \bigcup_{k=1}^{r} \tilde{\mathbf{a}}_k \times \tilde{b}_k \qquad (3.47)$$

Taking into account computational aspects, the Cartesian products $\tilde{\mathbf{R}}_k$, $k = 1, 2, \ldots, r$ are frequently implemented by using min or algebraic product functions, yielding two commonly used *operation rules:*

- *Mamdani's mini-operation rule:* $\mu_{\tilde{\mathbf{R}}_k}(\mathbf{u}, v) = \min(\mu_{\tilde{\mathbf{a}}_k}(\mathbf{u}), \mu_{\tilde{b}_k}(v))$;

- *Larsen's product operation rule:* $\mu_{\tilde{\mathbf{R}}_k}(\mathbf{u}, v) = \mu_{\tilde{\mathbf{a}}_k}(\mathbf{u}) \cdot \mu_{\tilde{b}_j}(v)$.

Detailed analysis of the influence of different fuzzy implication functions and union and intersection operators on quality of control can be found in, e.g., [251], [316], [204].

The inference mechanism is based on the sup-star compositional rule of inference:

$$\tilde{y} = \tilde{\mathbf{x}} \circ \tilde{\mathbf{R}} \qquad (3.48)$$

The rule is usually implemented using the sup-min or sup-product compositional operator. If this is the case, and if the union is implemented using the max function, the fuzzy control action $\tilde{y}$ can be expressed as:

$$\tilde{y} = \bigcup_{k=1}^{r} \tilde{\mathbf{x}} \circ \tilde{\mathbf{R}}_k = \bigcup_{k=1}^{r} \tilde{y}_k \qquad (3.49)$$

or, in terms of the degree of membership function, as:

$$\mu_{\tilde{y}}(v) = \max_{k \in 1, \ldots, r} \tilde{y}_k \qquad (3.50)$$

where $\tilde{y}_k = \tilde{\mathbf{x}} \circ \tilde{\mathbf{R}}_k$ is a local fuzzy control action inferred from the k−th rule. In terms of the degree of membership function, the local fuzzy control action is determined by:

$$\mu_{\tilde{y}_k}(v) = \begin{cases} \sup_{\mathbf{u}} \min \left[\mu_{\tilde{\mathbf{x}}}(\mathbf{u}), \mu_{\tilde{\mathbf{R}}_k}(\mathbf{u}, v) \right] & \text{in case of sup-min composition} \\ \sup_{\mathbf{u}} \left[\mu_{\tilde{\mathbf{x}}}(\mathbf{u}) \cdot \mu_{\tilde{\mathbf{R}}_{kj}}(\mathbf{u}, v_j) \right] & \text{in case of sup-product composition} \end{cases}$$
$$(3.51)$$

If the input $\tilde{\mathbf{x}}$ is a fuzzy singleton with membership function equal zero at all points except at the point $\mathbf{u}_0$ at which $\mu_{\tilde{\mathbf{x}}}(\mathbf{u}_0) = 1$, then both versions of the compositional rule of inference reduce to:

$$\mu_{\tilde{y}_k}(v) = \mu_{\tilde{\mathbf{R}}_k}(\mathbf{u}_0, v) \qquad (3.52)$$

In this way, a local fuzzy action is determined by the membership function:

$$\mu_{\tilde{y}_k}(v) = \begin{cases} \min \left[\mu_{\tilde{\mathbf{a}}_k}(\mathbf{u}_0), \mu_{\tilde{b}_k}(v) \right] & \text{in the case of Mamdani's rule} \\ \mu_{\tilde{\mathbf{a}}_k}(\mathbf{u}_0) \cdot \mu_{\tilde{b}_k}(v) & \text{in the case of Larsen's rule} \end{cases} \qquad (3.53)$$

The quantity $\mu_{\tilde{\mathbf{a}}_k}(\mathbf{u}_0)$ is referred to as the *firing strength* of the k-th rule and it represents a measure of the contribution of the k-th rule to the integral fuzzy control action. It is computed by applying the corresponding operation rule:

$$\mu_{\tilde{\mathbf{a}}_k}(\mathbf{u}_0) = \begin{cases} \min \left[\mu_{\tilde{a}_{k1}}(u_{10}), \mu_{\tilde{a}_{k2}}(u_{20}), \ldots, \mu_{\tilde{a}_{kn}}(u_{n0}) \right] & \text{for Mamdani's rule} \\ \mu_{\tilde{a}_{k1}}(u_{10}) \cdot \mu_{\tilde{a}_{k2}}(u_{20}) \cdot \;\cdots\; \cdot \mu_{\tilde{a}_{kn}}(u_{n0}) & \text{for Larsen's rule} \end{cases}$$
$$(3.54)$$

The mechanism of inference in these two methods is illustrated in Figure 3.8. The first method has the advantage that it enhances the contribution of the dominant rule, so that it is widely used in fuzzy applications. On the other hand, the second method has the advantage that it preserves the contribution of all rules to the control action.

The obtained fuzzy control action $\tilde{y}$ (or a set of local control actions $\tilde{y}_k$, $k = 1, 2, \ldots, r$) is transferred to the action, i.e., defuzzification interface, where the actual crisp control signal is generated.

3.5   Action interface

Degree of a membership function of fuzzy control action can be interpreted as a distribution of the possibility $\mu_{\tilde{y}}(v)$ to achieve the control goal by the signal v. The aim of the action interface is to generate such control signals that will best represent the possibility distribution of the inferred fuzzy action. The strategies frequently employed by the action interface are:

- *The mean-of-maximum method.* With this strategy, control action is derived as a mean value of all the points v at which the membership function of the fuzzy control reaches the global maximum $M = \max_v \left[\mu_{\tilde{y}}(v) \right]$. In the case of a discrete universe $V = \{\overline{v}_1, \overline{v}_2, \ldots, \overline{v}_q\}$, the control y is computed as:

$$y = \frac{1}{|L|} \sum_{l \in L} \overline{v}_l \qquad (3.55)$$

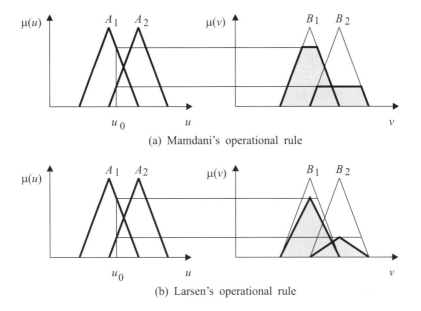

(a) Mamdani's operational rule

(b) Larsen's operational rule

Figure 3.8. Operational rules

where L is a set of all indices for which the grade membership function reaches maximum, i.e., $L = \{l : \mu_{\tilde{y}}(\overline{y}_l) = M\}$.

- *The centre-of-mass method.* With this method, control is generated as the centre of mass of the possibility distribution of fuzzy control. In the case of the discrete universe,

$$
y = \frac{\displaystyle\sum_{l=1}^{q} \mu_{\tilde{y}}(\overline{v}_l) \cdot \overline{v}_l}{\displaystyle\sum_{l=1}^{q} \mu_{\tilde{y}}(\overline{v}_l)}.
\tag{3.56}
$$

the analytical indices that could serve as guides in selecting the preferred method are not known. However, experiments conducted by several authors, e.g., [232], have shown that the mean-of-maximum method yields a better transient performance while the centre-of-mass method yields a better steady-state performance.

A slightly different scheme is employed with Sugeno-type rules, where fuzzy logic is employed only to describe conditions for the application of the rule, whereas the control actions are fuzzy singletons, i.e., the control signals in a classical sense:

$$R_k : \text{ if } X_1 \text{ is } A_{k1} \text{ and } \ldots \text{ and } X_n \text{ is } A_{kn} \text{ then } y_k = g_{k0} + \sum_{i=1}^{n} g_{ki} x_i$$

Here, it is natural to employ the firing strengths $\mu_{\tilde{\mathbf{a}}_k}(\mathbf{u}_0)$ of the rules to directly determine the crisp control signal y. A standard technique is to generate the aggregate signal as a weighted average of local controls, where the firing strengths are used as weighting factors:

$$y = \frac{\displaystyle\sum_{k=1}^{r} \mu_{\tilde{\mathbf{a}}_k}(\mathbf{u}_0) \cdot y_k}{\displaystyle\sum_{k=1}^{r} \mu_{\tilde{\mathbf{a}}_k}(\mathbf{u}_0)} \tag{3.57}$$

4.   Direct applications

The control rules frequently used in fuzzy controllers are of the type:

$$R_k : \text{ if } (E \text{ is } A_k \text{ and } \Delta E \text{ is } B_k) \text{ then } U = C_k$$

where E represents the value of the error e, ΔE represents the error change Δe between successive operation cycles of the controller, and U represents the fuzzy control action that is transferred to the action interface, which in turn generates the control signal u. For the operation of such a controller, of special importance are *normalizing gains* that are effectively applied within normalization that takes place during the conversion between actual signal values and their fuzzy representations. Normalized values of the signals can be represented as:

$$\begin{align}
e' &= e/E_{\max} = G_E \cdot e \tag{3.58}\\
\Delta e' &= \Delta e/\Delta E_{\max} = G_{\Delta E} \cdot \Delta e \tag{3.59}\\
u' &= u/U_{\max} = u/G_U \tag{3.60}
\end{align}$$

where $E_{\max}$, $\Delta E_{\max}$, $U_{\max}$ are maximum values of the signals, e', $\Delta e'$, u' are the corresponding normalized values, and G_E, $G_{\Delta E}$, G_U denote the normalizing gains. If the operation of the fuzzy controller is regarded as a nonlinear mapping from e and Δe to u, e.g.

$$u = G_U \cdot f(G_E \cdot e, G_{\Delta E} \cdot \Delta e) \tag{3.61}$$

then it is obvious that the fuzzy controller is actually a nonlinear PD controller. Conversely, the conventional PD controller can be considered as a special case of fuzzy PD controller for which the consequent of the rule is:

$$u = K_p e + K_d \Delta e \tag{3.62}$$

Analogously, a fuzzy controller with control rules of the type:

$$R_k : \text{if } (E \text{ is } A_k \text{ and } \Delta E \text{ is } B_k) \text{ then } \Delta U = C_k$$

where ΔU represents the change in control output, is actually a nonlinear PI controller. This analogy with conventional controllers is often employed in the synthesis and adjustments of fuzzy control.

5.   Synthesis of fuzzy controller - example from robotics

The control task of robot trajectory tracking is considered in the process of fuzzy controller synthesis. In order to simplify the example, only a manipulation robot with one DOF is presented. The control aim is that the robot end-effector tracks the desired trajectory with high accuracy. For this purpose it is necessary to acquire information about the positional tracking error e and the velocity tracking error Δe. The values of e and ΔE are inputs to the fuzzy controller, while its output is the control signal u. The information about positional and velocity tracking errors are taken from the robot sensors.

Each input and output variable is represented by five linguistic variables: *big negative (BN), small negative (SN), zero (Z), small positive (SP)* and *big positive (BP)*. This distribution is based on experience. Each of the input-output variables has its own physical range, which must be scaled on the appropriate number of quantization levels. These physical values are put in the frame of nine quantization levels $[-4, -3, -2, -1, -, 1, 2, 3, 4]$. Each of five linguistic values is defined by a fuzzy set with a Gaussian membership function (Figure 3.9):

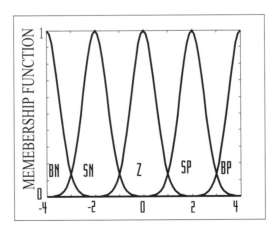

Figure 3.9.   Membership functions for 5 linguistic sets

$$\mu_{BIG\ NEGATIVE}(u) = e^{-0.5\left(\frac{u+4}{0.5}\right)^2}$$

$$\mu_{SMALL\ NEGATIVE}(u) = e^{-0.5(\frac{u+2}{0.5})^2}$$

$$\mu_{ZERO}(u) = e^{-0.5(\frac{u}{0.5})^2}$$

$$\mu_{SMALL\ POSITIVE}(u) = e^{-0.5(\frac{u-2}{0.5})^2}$$

$$\mu_{BIG\ POSITIVE}(u) = e^{-0.5(\frac{u-4}{0.5})^2}$$

The rule base consists of the following six generalized "IF-THEN" rules that are based on the knowledge and expert experience:

$RULE_1$: IF e is BP and Δe has any value THEN u is BP.
$RULE_2$: IF e is SP and Δe is SP or Z THEN u is SP.
$RULE_3$: IF e is Z and Δe is SP THEN u is Z.
$RULE_4$: IF e is Z and Δe is SN THEN u is SN.
$RULE_5$: IF e is SN and Δe is SN THEN u is SN.
$RULE_6$: IF e is BN and Δe has any value THEN u is BN.

These fuzzy rules can be easily interpreted using Table 3.1:

Table 3.1.   Rule base for fuzzy controller

$e/\Delta e$	BN	SN	Z	SP	BP
BN	BN				BP
SN	BN	SN	SN		BP
Z	BN			SP	BP
SP	BN		Z	SP	BP
BP	BN				BP

In the inference process, the Mamdani implication and min-max reasoning operator are used. In the particular case when, at the input to the fuzzy robot controller, there exist signals of position and velocity errors as fuzzy singleton values $e^* = 1$ and $\Delta e^* = 2.5$, then, based on the fuzzy inference mechanism, the rules $RULE_2$ and $RULE_3$ are fired (Figure 3.10). Using a defuzzifaction process based on the gravity method, the output value $u^* = 1$ is formed. This procedure must be repeated in every sampling period of the robot fuzzy controller.

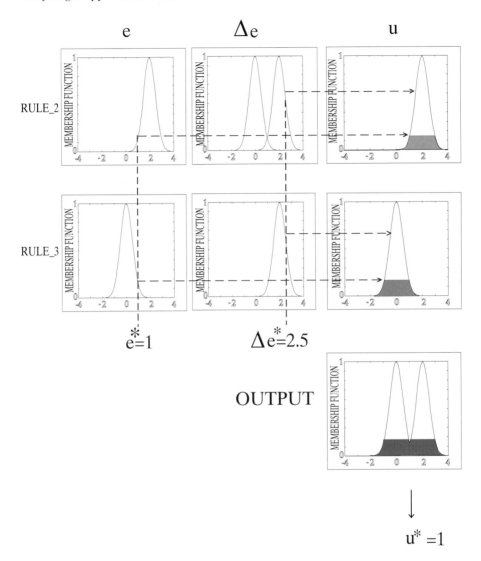

Figure 3.10. Inference engine and defuzzification of robot controller

6. Fuzzy algorithms in robotics

Considerable attention has been paid to the application of fuzzy control techniques to modern robotic systems. The algorithms of fuzzy logic in robotics can be classified according to the particular robotic tasks and hierarchical control levels [364],[347], [319]:

a) *Tracking control of robot trajectories* [232], [270], [255], [117], [298], [216], [348], [277], [161], [336], [292], [394] - executive (adaptation) control level

b) *Force control and hybrid position/force control for robotic contact tasks* [350],[301],[111],[162], [303],[221],[129] - executive (adaptation) control level

c) *Motion control of mobile robots and autonomous unmanned vehicles* [142], [175], [141], [332], [7], [382], [391], [50], [120], [227], [100], [346], [286], [99], [374] ,[265] , [6], [252], [228], [67], [299] - executive (adaptation), tactical (skill) and strategical (learning) control level

d) *Control of locomotion, rehabilitation and special robotic systems* [63], [343], [320], [234], [409], [363], [388], [146], [193] - all control levels

d) *Trajectory generation and task planning* [65], [318], [312], [23], [387] - tactical (skill and learning) control level

e) *Use of fuzzy logic for expert robot control and for modelling/supervising of robotic systems* - strategical (learning) level [62], [217]

f) *Fuzzy logic for vision systems and other sensor robotic systems* [142], [332], [240], [19], [58].

g) *Fuzzy logic approach to control cooperative manipulation* - [101] - executive control level

h) *Fuzzy logic approach to robotic teleoperation* - [184] - executive control level

There are three principal architectures for positional tracking control of manipulation robots. A simple application of the first architecture (fuzzy architecture as part of a closed-loop system with feedback) has been described, in [216]. A classical fuzzy controller is applied without using the mathematical model of robot with one DOF (Figure 3.11). Control of the robot system is performed using the signals of positional and velocity error with two levels of quantization and a rule base with the min-max method of fuzzy inference. In comparison with the conventional PI controller and MRAC (Model Reference Adaptive Controller), the proposed fuzzy controller shows better results in transient regimes, as well as in a steady-state regime. However, the drawback of the proposed fuzzy structure is the lack of analytical tools for the control synthesis.

A second group of fuzzy controllers in robotics is represented by the fuzzy control architecture for tuning parameters of conventional PID robot controllers [348], [277], [161], [230], [292]. These control structures can be with tuning of robot controllers with or without the mathematical model of the robotic systems (Figure 3.12). The latter hybrid approach shows great potential for accuracy of tracking, because the tuning of PID controllers is achieved by fuzzy structure using the practical stability of the conventional controller with robotic subsystem models [161].

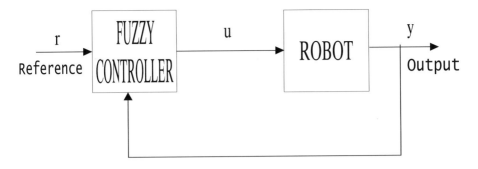

Figure 3.11. Fuzzy robot control structure as part of closed loop systems with feedback

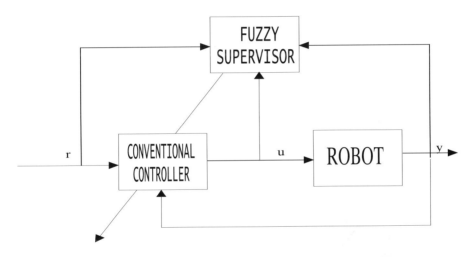

Figure 3.12. Tuning fuzzy robot control architecture

A first example of the application of fuzzy control to robotic manipulators was given by Mandic, Scharf and Mamdani [232] using the third architecture - self-organizing fuzzy controller. In their paper from 1985, the authors described a series of experiments with a two-DOF robot controlled via two independent self-organizing controllers. Both controllers are of the same structure (see Figure 3.13) that consists of two levels.

The lower level is a usual fuzzy controller with control rules, whereas the upper level is a system that realizes the mechanism of automatic learning, i.e., the generation and modification of the rules at the lower level. Control rules employed on the lower level were of the type:

$$R_k : \text{if } (E \text{ is } A_k \text{ and } \Delta E \text{ is } B_k) \text{ then } U = C_k$$

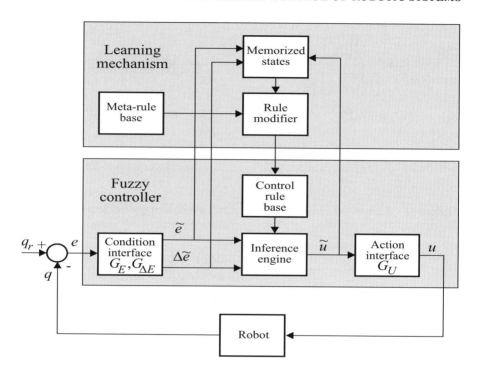

Figure 3.13.   Self-organizing controller

where E represents the joint position error $e = q - q_0$, ΔE represents the error change, and U represents the control action. The upper level is responsible for evaluation of the controller performance and modification of control rule base. The evaluation of performance is achieved using a production system whose structure is identical to the basic fuzzy controller. The performance is evaluated using a local criterion that roughly expresses the difference P between the actual and desired system response. Evaluation criteria are expressed using a set of meta-rules of the form:

$$\Pi_k : \text{if } (E \text{ is } V_{Ek} \text{ and } \Delta E \text{ is } V_{Ek}) \text{ then } P = V_{Pk}$$

In these rules, parameters of the primary fuzzy set $V_{Pk} = \text{ZERO}$ define a tolerance range for the system response, whereas the values different from ZERO imply the desired degree of correction. Remarkably, the rules thus defined depend to a very small extent on the controlled process and they really express the tolerable errors and degree of acceptability of the errors. If the base of meta-rules is represented by the fuzzy relation Π, then the output of the evaluator is a nonlinear function:

$$p = \pi(e, \Delta e) \tag{3.63}$$

If a precise model of the controlled system were available, then a necessary correction in control Δu could be in principle calculated from the known index p. Since the use of a model is always accompanied by inaccuracies, the self-organizing controller instead performs a modification of the control rules that is based on simplified assumptions that (1) the current system performance index $p(t)$ is a consequence of the control $u(t - nT)$, generated n operation cycles prior to the current time instant t and (2) the necessary correction in control $\Delta u(t - nT)$ is proportional to $p(t)$:

$$r(t) = \Delta u(t - nT) = \lambda \cdot p(t) \tag{3.64}$$

In other words, it is accepted that the corrections in control are not one hundred per cent accurate and that by this the learning process is slower.

Modifications in the rule base are achieved using fuzzy set operations. Namely, the rule base at the time instant t can be represented as the union:

$$\tilde{R}(t) = \bigcup_{k=1}^{r(t)} \tilde{e}_k \times \Delta\tilde{e}_k \times \tilde{u}_k \tag{3.65}$$

If the function transforming a value x into the fuzzy singleton $\tilde{x} = F\{x\}$ is denoted by $f\{\cdot\}$, then the control generated at the time instant $t - nT$ may be regarded as a value that corresponds to the fuzzy implication:

$$\tilde{R}'(t) = f_E\{e(t - nT)\} \times f_{\Delta E}\{\Delta e(t - nT)\} \times f_U\{u(t - nT)\} \tag{3.66}$$

whereas the desired control at the current time instant is regarded as a value corresponding to the implication:

$$\tilde{R}''(t) = f_E\{e(t-nT)\} \times f_{\Delta E}\{\Delta e(t-nT)\} \times f_U\{u(t-nT) + \lambda \cdot p(t)\} \tag{3.67}$$

Now, the problem of modification of control rules can be expressed as a problem of substitution of the implication $\tilde{R}'(t)$ by the implication $\tilde{R}''(t)$. One of the ways to achieve this is to describe the substitution by the expression:

$$\tilde{R}(t + T) = [\tilde{R}(t) \text{ and not } \tilde{R}'(t)] \text{ or } \tilde{R}''(t) \tag{3.68}$$

or, equivalently:

$$\tilde{R}(t + T) = [\tilde{R}(t) \cap \overline{\tilde{R}'(t)}] \cup \tilde{R}''(t) \tag{3.69}$$

Direct application of this formula would lead to an exponential growth of the number of rules. Therefore, an approximate method is used where only single rule is modified at the time. The modified rule is the dominant rule, i.e., the rule $\tilde{R}_k = \tilde{e}_k \times \Delta\tilde{e}_k \times \tilde{u}_k$ for which the heights of the intersections $\tilde{e}_k \cap f_E\{e(t - nT)\}$ and $\Delta\tilde{e}_k \cap f_{\Delta E}\{\Delta e(t - nT)\}$ are at least equal 0.5. Once the dominant rule is identified, its old fuzzy action $\tilde{u}_k$ is replaced by the action $f_U\{u(t - nT) + \lambda \cdot p(t)\}$.

In the experiments by Mandic et al. it was demonstrated on a real robot that a self-organizing controller, after only a few adaptation cycles, was capable of attaining the steady performance that was completely comparable to that of a conventional PID controller. Similar results achieved by Tanscheit and Scharf [336], who described several experiments with the self-organizing controllers applied to control a second-order linear system, representing the transfer function of a D.C. motor with variable load (variable load was represented by the variable moment of inertia).

In these and other initial works, in an attempt to attain direct control of manipulation robots by fuzzy controllers, two main problems have arisen. The first is manifested by the lack of analytical tools for control synthesis, i.e., the selection of the parameters of the fuzzy controllers (or initial values of the parameters in case of self-organizing controllers). Second, ordinary fuzzy controllers have attained the performance similar to, or slightly better than simple PID schemes. Therefore, it may be expected that a direct application of fuzzy controllers will not yield satisfactory performance in more complex robotic tasks, such as tracking of fast trajectories. The appearance of these problems can be partially explained by the fact that the early works were primarily concentrated on demonstration of the ability of fuzzy logic-based methods to effectively master nonlinear control problems without the need for exact mathematical modelling of the controlled system. For this reason, the role of a priori available mathematical knowledge (in situations where the system dynamics is deterministic), as well as the established model-based control techniques, were somewhat overshadowed.

The problem of merging fuzzy logic-based control with analytic methodologies to exploit the advantages of both approaches in real-time robot control was addressed by several authors. Lim and Hiyama [217] proposed a decentralized control strategy that incorporates a PI controller and a simple fuzzy logic controller. In their approach, the PI controller was used to enhance transient response and steady-state accuracy, whereas fuzzy control was aimed at enhancing damping of the overall system. A tighter connection between fuzzy and standard control methods was proposed by Tzafestas and Papanikolopoulos [348], who suggested to employ a two-level hierarchy in which a fuzzy logic-based expert system is used for fine tuning of low-level PID control. A similar approach was applied to robot control by Popovic and Shekhawat [277]. However, the two-level control hierarchy by itself does not actually solve the problem of weak performance in situations that are characterized by fast varying robot dynamics. In such cases, the knowledge of readily available mathematical model of robot dynamics cannot be ignored. Therefore, fuzzy logic-based control should not be viewed as a pure alternative to model-based robot control.

Instead, a combined approach is preferred, and it may yield superior control schemes to both simple model-based and fuzzy logic-based approaches.

The general idea behind the hybrid approach is in the utilization of a satisfactory approximation of the model of robot dynamics to decrease dynamic coupling between the robot joints and then in engaging the fuzzy logic-based heuristics as an effective tool for creating a nonlinear performance-driven PID control to handle the effects not covered by the approximate model. A similar concept was formulated by de Silva and MacFarlane [61], who proposed a three-level hierarchy for robot control. The proposed hierarchy consists of:

1 Conventional robot controller, i.e., a set of PID controllers closed around a fast decoupling controller. The important role of this controller is to ensure the decoupling and linearization needed to efficiently apply expert knowledge for tuning the PID controllers (in an idealized case, every joint subsystem would behave as an independent oscillator with PID control);

2 Intelligent pre-processor, i.e., a set of knowledge-based observers. Each observer is implemented as a fuzzy system and its outputs are the attributes of response at the corresponding joint (e.g., accuracy, oscillations, error convergence, divergence, and steady-state error);

3 Fuzzy tuner, i.e., a fuzzy controller that is used for tuning the gains of the PID controllers to the lowest level.

The authors have tested their approach by simulating a two-link manipulator with an assumption of idealized effectiveness of the low-level global nonlinear feedback. Therefore, the robot dynamics was approximated by a set of joint subsystems modelled as second-order systems with unknown acceleration-type disturbances.

The idea of a hybrid approach to robot control was elaborated in detail by Vukobratovic and Karan [357], [358], who employed fuzzy logic to express control policy and determined analytically the conditions for the values of fuzzy control parameters that ensure stability of the closed-loop robot control system. The authors have analyzed a hybrid design, which is an extension to the decentralized control strategy. The proposed controller consists of a set of subsystems, closed around individual robot joints, where each of the subsystems comprises two components: conventional model-based controller and fuzzy logic-based tuner (see Figure 3.14). Inputs to the $i-$th joint subsystem, $i = 1, \ldots, n$, where n is the number of actuated joints, are the nominal control signal u_{0i}, joint position error Δq_i, joint velocity error $\Delta \dot{q}_i$, and the integral error $\int \Delta q_i dt$. In cases where highly precise tracking of fast trajectory is necessary, an optional global feedback loop (full dynamic compensation) can be added. The global feedback is generated on the basis of the computed or measured deviation of the dynamic torque $\Delta\tau_i$ acting at the joint, and it is synthesized to ensure practical system stability [361].

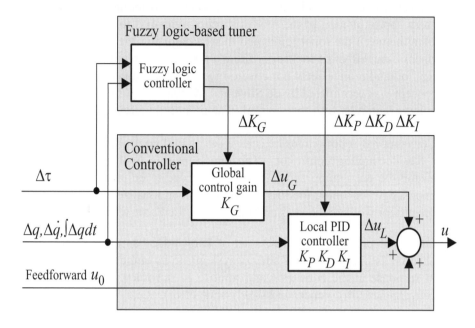

Figure 3.14. Hybrid control scheme

A further refinement consists of the introduction of the upper level to tune the gains of the PID controllers. The tuner is designed as a fuzzy controller that monitors joint response characteristics and modifies the gains to provide better responses for large deviations of the monitored quantities. Although its general structure permits one to construct sophisticated control rules for tuning gains, Vukobratovic and Karan have considered a simple decentralized scheme consisting of independent joint servo tuners, operating on the basis of the observed joint position error Δq, velocity error $\Delta\dot{q}$, and integral error $\int \Delta q dt$. A rather simple heuristics for synthesizing gain-tuning rules was used:

(a) If the observed errors are large and they do not show a significant tendency to decrease, the proportional gain is enlarged to speed-up error convergence.

(b) When the errors are small, the proportional gain is decreased to prevent resonance oscillations and attenuate undesired noise effects.

(c) If the errors are large, but the error convergence is satisfactory, the proportional gain is gradually decreased down to the value that is appropriate for small-error conditions.

(d) The values of derivative and integral gains are changed simultaneously with the changes in proportional gain so that the stability of the whole system is preserved. The actual values are derived from a stability analysis of the closed-loop system. Importantly, the readily available stability conditions for a fixed-gain controller were reused to determine the conditions on parameters of nonlinear fuzzy tuners that are sufficient for the overall system stability.

In spite of their simplicity, the rules resulted in significant improvements compared to that of a fixed-gain model-based controller. Simulation experiments on a real-scale 6-DOF industrial robot have shown that the resulting variable-gain controllers in many respects outperform constant-gain schemes. The most obvious advantage was the improvement in accuracy that was demonstrated in both positioning and trajectory tracking tasks. An important feature was that the accuracy improvement was not accompanied by degradation in other performance characteristics, such as energy consumption and maximum developed torques. Another significant aspect is the possibility of reducing the computational complexity of the nominal robot model (by employing approximate robot models) without the significant loss in control quality that was notable with fixed-gain control. Although issues related to sensitivity to parameter variations were not explicitly investigated, an improved robustness of the variable-gain controller is implied by the results obtained from the experiments with the approximate robot models.

For force control and hybrid position/force control in robotics, the basic principle of application of fuzzy logic is the same as in the case of position tracking control, except for the input level, where the space of input variables is extended with force tracking errors. The conventional force controllers give satisfactory results in the case when the parameters of the robot environment are known and fixed. However, this assumption in most cases is not fulfilled, so that it is necessary to adapt the control structure and parameters of robot controllers with respect to the changing environment. One of the techniques for solving these problems is the application of fuzzy logic [127],[128], [340],[129],[130]. In the process of control synthesis, there are many restrictive assumptions connected with the availability of force derivation and the possibility to measure the velocity of the robot end-effector, and these assumptions are effective only when the estimations of environment parameters are close to their real values. However, in practical application, information about force derivation contains measured noise, and the accuracy of position and velocity of the end-effector is not sufficient for satisfactory force control. Also, the range of parameters of the working environment can be very wide because, for example, the stiffness coefficient can vary in a large span, depending on whether the environment is soft or hard. The purpose of the fuzzy scheme of force control [340](Figure 3.15) is to achieve exact force regulation when the stiffness of the working environment is an unknown with great variation, using no restrictive assumptions.

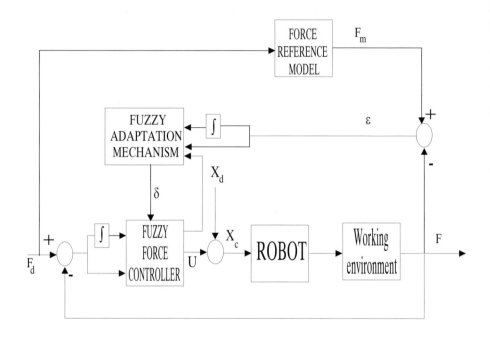

Figure 3.15.   Fuzzy scheme of force control

The proposed adaptive fuzzy scheme for control of the force F has three main components: force reference model, which describes the desired behaviour of the control scheme according to the force component F_m, main force fuzzy controller which calculates the adaptation U in the position control loop, and the algorithm of fuzzy learning and adaptation (command δ), which modifies the main fuzzy force controller based on the difference between the real and desired force. A common way of defining the force reference model is the synthesis of a second-order linear system in state space with a specified damping factor and the natural frequency of the system. The main fuzzy controller is a fuzzy PI controller with two input values (force error and integral force error) and one output value, which represents the adaptation of the position control loop. The structure of the fuzzy controller is defined by simple fuzzy rules with the membership functions of five fuzzy sets. The task of the fuzzy mechanism of adaptation and learning is to learn parameters of the working environment and modify the fuzzy force controller so that the system tracks the output of force reference model. In this case, inputs to the fuzzy mechanism are force error and integral force error, while output is a modification of the output of the basic fuzzy force controller. This modification is represented by a change of the membership function of the output fuzzy set with respect to the centre and width of the membership function. The efficiency of this approach is demonstrated by simulation studies with different parameters of the working environment.

Furthermore, fuzzy controllers can be applied as an additional part of the basic control algorithms in order to compensate for system uncertainties as in the case of the application of fuzzy controllers in a sliding mode regime [162], [129]. Another important application of fuzzy logic is to control the interphase between the position control and force control in order to avoid great values of force and defracture of the working object. [303]. The fuzzy logic is efficiently applied for the DSP control of robot hand fingers [221] to achieve the appropriate grasping force.

One of the earliest applications of fuzzy controllers for navigation of mobile robots and vehicles was described in [321]. The Takagi - Sugeno Fuzzy Controller is applied to the problem of car parking. The linguistic rules are given by modelling human action in the process of car parking. Since that time, there have been many research reports dealing with the application of fuzzy controllers in mobile and autonomous robotics [382],[31],[406],[120], [227], [100],[346],[374]. The main task of fuzzy controllers for autonomous mobile robots and vehicles is to ensure target steering and collision avoidance in the space of obstacles. Optimal control of autonomous mobile robots requires operation with imprecise information about the robot movement and working environment using information from different robot sensors (in most cases these are ultrasound and vision sensors). The input fuzzy sets represent the information about the robot distance from the obstacles, current velocity and angle of environment perception by camera, while the output fuzzy sets represent the decision about the change of speed, acceleration, and deviation angle based on the fuzzy inference mechanism. Sometimes, based on human experience, it is efficient to use distributed fuzzy control with separated direction control and velocity control [120]. In a specific algorithm used in mobile robotics [227] fuzzy logic was applied as a part of look-ahead control for target tracking of moving objects (mobile robot tracks another mobile robot). It is interesting that this fuzzy controller has as input the predicted position error and velocity error based on Gray Prediction Theory. Compared with conventional fuzzy controller and conventional target tracking strategies, this prediction-based fuzzy controller can update the system dynamic model parameters according to limited information from uncertain sensory measurement. In this case, convergence rate of the prediction-based fuzzy controller is much better than that of conventional fuzzy controller. Fuzzy logic-based navigation solutions are very convenient for the navigation of autonomous robotized underwater vehicles [100] because sonar data is unreliable, mathematical models of the environment and vehicle are usually not available, and the available navigation expertise is rather poor. Hence, using a three-layer structure of fuzzy control [100] it is possible to achieve accurate and collision-free navigation. Fuzzy logic was used for sensor fusion in order to calculate the collision possibility, together with calculation of the necessary corrections for the thrusters of autonomous vehicles. In [382], a sensor-based fuzzy algorithm is developed for navigating a mobile robot in a two-dimensional unknown environment with stationary polygonal obstacles. In this algorithm, the robot initially scans and measures by the sensors the angles and distances of all the visible vertices of the obstacles. The priority of each visible vertex is determined by feeding the readings of the sensors to a suitable set of priority fuzzy rules. When the priorities of all visible vertices are found, the navigation algorithm selects the highest one and guides the robot to the corresponding vertex. Ward and Zelinsky [374] proposed the approach which is based on learning a fuzzy associative map between the sonar sensors and the vehicle. The results demonstrate that the use of multiple fuzzy associative map matrices can significantly improve the robotic perception and behaviour by enabling the environment to be resolved with more accuracy.

In [67], a two-layer fuzzy logic controller was designed for 2-D autonomous navigation of a skid steering vehicle in an obstacle-filled environment. The derived and implemented planner is a two-layer fuzzy logic based controller that provides purely reactive behaviour of the vehicle moving in a 2-D obstacle-filled environment, with input readings from a ring of 24 sonar sensors and angle errors, and outputs the updated rotational and translational velocities of the vehicle. In the first layer, there are four fuzzy logic controllers responsible for obstacle detection and calculation of the collision possibilities in the four main directions, front, back, left, right. The four controllers receive as inputs the sonar sensor data and return as output the collision possibility in the directions front, back, left and right. The possibilities calculated in the first layer are the input to the second layer along with the angle error (the difference between the robot heading angle and the desired target angle), and the output is the updated vehicle's translational and rotational speed. The second layer fuzzy controller receives as inputs the four collision possibilities in the four directions and the angle error, and outputs the translational velocity, which is responsible for moving the vehicle backward or forward, and the rotational speed, which is responsible for the vehicle rotation. The angle error represents the difference between the robot heading angle and the desired angle the robot should have in order to reach its target. The linguistic variables that represent the angle error are: $Backwards_1$, $Hard_{Left}$, $Left$, $Ahead$, $Right$, $Hard_{Right}$, $Backwards_2$. The two layer controller applicability and implementation is demonstrated through experimental results and case studies performed on a real mobile robot.

Seraji and Howard [299] present a new strategy for behaviour-based navigation of field mobile robots on challenging terrain, using a perception-based linguistic framework and a novel measure of terrain traversability. The main advantages of the fuzzy navigation strategy lie in the ability to extract heuristic rules from human experience, and to obviate the need for an analytical model of the process. Also, a key feature of the proposed approach is real-time assessment of terrain characteristics and incorporation of this information in the robot navigation strategy. The proposed approach is highly robust in coping with the uncertainty and imprecision that are inherent in the sensing and perception of natural environments. Three terrain characteristics that strongly affect its traversability, namely, roughness, slope, and discontinuity, are extracted from video images obtained by on-board cameras. This traversability data is used to infer, in real time, the terrain Fuzzy Rule-Based Traversability Index, which succinctly quantifies the ease of traversal of the regional terrain by the mobile robot. A new traverse-terrain behaviour is introduced that uses the regional traversability index to guide the robot to the safest and the most traversable terrain region. The regional traverse-terrain behaviour is complemented by two other behaviours, local avoid-obstacle and global seek-goal. The recommendations of these three behaviours are integrated through adjustable weighting factors to generate the final motion command for the robot. Multiple fuzzy navigation behaviours are combined into a unified strategy, together with smooth interpolation between the behaviours to avoid abrupt and discontinuous transitions. The weighting factors are adjusted automatically, based on the situational context of the robot. The final motion commands are computed using the Centroid defuzzification method. The terrain assessment and robot navigation algorithms have been implemented on a Pioneer commercial robot and field-test studies are being conducted. The field test studies reported demonstrate that the mobile robot possesses intelligent decision-making capabilities that are brought to bear in negotiating hazardous terrain conditions during robot motion.

The procedure of robot task planning and trajectory generation is based on the synthesis of internal robot models of the robot environment. The problem is related to environment complexity, because the environment can not be represented in detail by mathematical models. Many uncertainties are present in the robotic systems due to the unreliability and incompatibility of the robot sensors and the end-effector, and the impossibility to represent the object, locate object, or the impossibility to achieve appropriate action with sufficient accuracy. In this case, fuzzy logic gives the methodology for designing functional mapping between the state space of the robot and state space of the external robot environment. The previously defined fuzzy relation models represent the uncertainties of the external robot environment. Based on example of planning of robot movement velocity through appropriate passage, this is shown using fuzzy relation between the internal robot model and model of the external robot environment. The rows in Table 3.2 denote the robot internal state value using linguistic terms: "very wide (VW)", "wide (S)","narrow (U)". The columns in the table represent the degree of membership between the internal robot model and external model of the real robot environment.

Table 3.2. Fuzzy relation for passage width

Passage width		Robot environment		
	0.5m	1m	1.5m	2m
Very Wide	0	0.1	0.25	1
Wide	0	0.3	0.5	1
Narrow	1	0.7	0	0

As example of exact interpretation of the above table it is taken that the linguistic variable "Wide" (W) belongs to the fuzzy set "1.5m" with the degree of membership 0.5. In a similar way, the robot velocity is defined as the second important value for robot control using Table 3.3.

Table 3.3. Fuzzy relation for robot velocity

Robot velocity			Real velocity		
	0.2m/s	0.5m/s	0.8m/s	1m/s	1.5m/s
Big	0	0	0.2	0.8	1
Average	0.2	0.7	1	0.8	0.1
Small	1	0.9	0.2	0	0

The aim of the robot controller is to plan the robot velocity according to specific input information about the passage width. Using expert knowledge about the required task, the following rules are defined:

A) IF the passage witdh is "Very Wide", THEN the robot velocity must be "Big"
B) IF the passage width is "Wide", THEN the robot velocity must be "Big"
C) IF the passage width is "Narrow", THEN the robot velocity must be "Small"

In the process of rule interpretation using fuzzy implications, Table 3.4 is obtained, while the same rules can be interpreted using conventional logic according to Table 3.5

Table 3.4. Fuzzy relations of the control algorithm

				Real velocity			
		0.2m/s	0.5m/s	0.8m/s	1m/s	1.5m/s	
Passage width	0.5m	1	0.9	0.2	0	0	
	1m	0.7	0.7	0.2	0.3	0.3	
	1.5m	0	0	0.2	0.5	0.5	
	2m	0	0	0.2	0.8	1	

Table 3.5. Production rules

PLANNING RULES			Real velocity		
		Big	Average	Small	
	Very Wide	1	0	0	
Passage width	Wide	1	0	0	
	Narrow	0	0	1	

In the application of fuzzy algorithms using modern robotic sensors, because of the problem complexity and system uncertainties, it is necessary to use intelligent signal processing. One of efficient applications of fuzzy logic for sensor systems is based on the application of ultrasound sensors for measuring robot distance from the surrounding obstacles [240],[19]. Another important application is the visual servoing of a robot manipulator in the case of target identification and grasping of spherical objects [58]. Because of the problem of calibration errors or the impossibility to measure exactly distance from the camera to the target, the task of visual servoing needs some alternative techniques. One of them is fuzzy logic, whereby this technique is used to easily detect situation-action mapping. In this way the fuzzy logic controller can incorporate the key components of reactive controllers and formal reasoning on uncertain information of the vision system.

Some researchers used fuzzy logic [363], [409], [388], [146] as the methodology for biped gait synthesis and control of biped walking. Fuzzy logic was used mainly as part of control systems on the executive control level, for generating and tuning PID gains, fuzzy control supervising, direct fuzzy control by supervised and reinforcement error signals.

In [363], fuzzy logic is applied on the level of local control for the tuning of local PID gains, while the complete control structure includes nominal feedforward control (based on the biped's dynamics model), too. It has been shown that the aggregation-decomposition method for stability analysis of the overall biped system is applicable in cases when the local subsystems are stabilized with fuzzy regulators. For the synthesis of fuzzy regulators, a method of parallel distributed compensation was utilized.

The fuzzy logic approach is also applied to the complex problem of simultaneous position and internal force control in multiple cooperative manipulator systems [101]. The controllers objectives are to track a predefined desired position and orientation of the payload, while controlling the internal forces of the closed-chain system, and make them converge to their predefined desired values. The controller has to attain these objectives in the face of significant parametric and modelling uncertainties and external disturbances and with partial or no prior knowledge of the system's dynamics. A decentralized adaptive hybrid intelligent control scheme is proposed here. The controller consists of an innovative hierarchical structure with two types of adaptive control modules: a conventional one and a adaptive fuzzy control, which makes use of an online adaptation scheme to fully assess and approximate the overall system dynamics, starting from partial or no a priori knowledge of it. In addition, an efficient fuzzy rule reduction scheme is suggested to drastically reduce the computational complexity of the adaptive fuzzy control module of the controller. The controller makes use of a multi-input multi-output fuzzy logic engine and a systematic online adaptation mechanism. Unlike conventional adaptive controllers, the proposed one does not require a precise model of the system dynamics. Numerical simulations are carried out to assess the performance of the hybrid intelligent adaptive controller when compared to a conventional one and to illustrate its robustness and ability to meet the predefined goals for the different levels of modelling uncertainties and external disturbance. It has been shown that the proposed controller is robust in the face of a substantial amount of parametric and modelling uncertainties and external disturbances with varying intensity levels. The proposed controller is also generic as it is independent of the robotic system in use and can be applied to almost any cooperative manipulator system, with possibly a few minor tune-ups.

The problem of biped gait synthesis using reinforcement learning with fuzzy evaluative feedback is considered in [409]. As first, an initial gait is generated from fuzzy rules using the human intuitive balancing scheme. Simulation studies showed that the fuzzy gait synthesizer can only roughly track the desired trajectory. A disadvantage of the proposed method is the lack of practical training data. In this case there are no numerical feedback teaching signals, only evaluative the feedback signal exists (failure or success), exactly when the biped robot falls (or almost falls) down. Hence, it is a typical reinforcement learning problem. The dynamic balance knowledge is accumulated through reinforcement learning constantly improving the gait during the walk. Exactly, it is fuzzy reinforcement learning that uses fuzzy critical signals. For a human biped walk, it is typical to use linguistic critical signals such as "near-fall-down", "almost-success", "slower", "faster", etc. In this case, the gait synthesizer with reinforcement learning is based on a modified GARIC (Generalized Approximate Reasoning for Intelligent Control) method. This architecture of gait synthesizer consists of three components: action selection network (ASN), action evaluation network (AEN), and stochastic action modifier (SAM) (Figure 3.16). The ASN maps a state vector into a recommended action using fuzzy inference. The training of an ASN is achieved as with standard neural networks using error signal of external reinforcement. The AEN maps a state vector and a failure signal into a scalar score, which indicates the state goodness. It is also used to produce internal reinforcement. The SAM uses both recommended action and internal reinforcement to produce a desired gait for the biped. The reinforcement signal is generated based on the difference between the desired ZMP and real ZMP in the x-y plane. In all cases, this control structure includes on-line adaptation of gait synthesizer and local PID regulators. The approach is verified using simulation experiments. In the simulation studies, only even terrain for biped walking is considered, hence the approach should be verified for irregular and sloped terrain. $Xzmp, Yzmp$ are the ZMP coordinates;

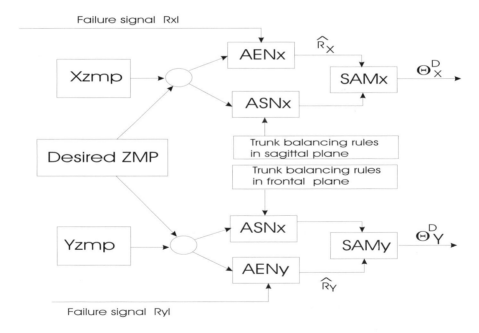

Figure 3.16. Architecture of the reinforcement learning based gait synthesizer

$\theta^d_{zmp}, \theta^d_{zmp}$ are the desired joint angles of the biped gait.

In [388], a conventional fuzzy controller for position/force control of the robot leg is proposed and experimentally verified. This intelligent walking strategy is specially intended for walking on rough terrain.

A main problem in the synthesis of fuzzy control algorithms for biped robots remains the inclusion of a dynamics model and learning capabilities in order to obtain exact tracking of biped trajectories as well as the steps with greater speed, preserving the dynamic stability of the biped gait.

In addition to the mentioned approaches to the application of fuzzy logic in robotics as the elements of control systems, an important issue in pertinent research is the modelling and identification of robotic systems using dynamic fuzzy logic systems [205].

Chapter 4

GENETIC ALGORITHMS IN ROBOTICS

1. Introduction

Today, contemporary computer methods inspired by biological evolution are grouped under the field called evolutionary computation. Evolutionary computation is the name given to a collection of algorithms based on the evolution of a population toward a solution of a certain problem. These algorithms can be used successfully in many applications requiring the optimization of a certain multi-dimensional function. The population of possible solutions evolves from one generation to the next, ultimately arriving at a satisfactory solution to the problem. These algorithms differ in the way a new population is generated from the present one, and in the way the members are represented within the algorithm. The three main elements of evolutionary computation are: 1) *evolution algorithms (EA)*; 2) *genetic programming (GP)*; 3) *genetic algorithms (GA)*. Each of these three techniques mimics the processes observed in natural evolution, and provides efficient search engines by evolving populations of candidate solutions to a given problem.

Genetic programming is a special implementation of GAs. It uses hierarchical genetic material that is not limited in size. The members of a population or chromosomes, are tree-structured programs and the genetic operators work on the branches of these trees. The structures generally represent computer programs written in LISP.

Evolutionary algorithms do not require separation between a recombination and an evaluation space. The genetic operators work directly on the actual structure. The structures used in EAs are representations that are problem dependent and more natural for the task than the general representations used in GAs.

Genetic programming is currently experiencing a dramatic increase in popularity. GP have been used to solve problems in various domains, including process control, data analysis, and computer modelling. Although at the present time the complexity of the problems being solved with GP lags behind the complexity of applications of various other evolutionary computing algorithms, the technique is promising. Because GP actually manipulates entire computer programs, the technique can potentially produce effective solutions to very large-scale problems. To reach its full potential, GP will likely require dramatic improvements in computer hardware.

GAs are generally thought to be the most prominent technique in the field of evolutionary computation. These algorithms utilize an iterative approach to solving search problems. GAs represent global search algorithms based on the mechanism of natural selection and natural genetics. This mechanism is based on a Darwinian-type survival-of-the-fittest strategy with reproduction, where stronger individuals in the population have a higher chance of creating an offspring. They are population-based search techniques that rely on the information contained in a broad group of candidate solutions to solve the problem at hand. This population-based search distinguishes GAs from traditional point-by-point search routines. It is not a gradient search technique, because they combine survival of the fittest among string structures (binary or nonbinary type) with a structured, yet randomized, information exchange. This randomized search can be very efficient and effective in avoiding local minima. The strength of GAs lies in their ability to implicitly identify the inviting properties associated with potential solutions, and to produce subsequent populations of candidate solutions, which contain new combinations of these fertile characteristics as derived from candidate solutions in preceding populations. Furthermore, GA is not considered to be a mathematically-guided algorithm. It is merely a stochastic, discrete event and a nonlinear process, which gives the optima containing the best elements of previous generations. GAs are well grounded in theory, and empirical studies demonstrate that in some problems, they consistently outperform more traditional search routines.

This approach to solving search problems may seem unusual at first consideration. However, GAs demonstrate at least three characteristics that allow them to perform efficiently in difficult search spaces. First, GAs consider a population of solutions simultaneously. Thus, they tend to maintain a global perspective due to their parallel consideration of solutions, whereas traditional search routines tend to perform local searches in the point-by-point manner, in which they traverse a search space. Second, GAs do not rely on derivative information. Thus, they tend not to be fooled by multi-modal search problems that perplex the derivative-based techniques so popular in the field of optimization. Third, GAs do not depend on continuity of the search space. Thus, they have been used to solve problems that have stymied more traditional search routines which can break down when applied in discontinuous search spaces.

GAs code the requisite information for a solution to a given problem in strings called chromosomes. More precisely, the GA operates on a space of *genotypes* (chromosomes) - the representatives of the corresponding elements in the search space. The former are usually referred to as *phenotypes*. Each chromosome can be decoded according to a user-defined mapping function, yielding specific values for each of the important parameters being sought. Coding schemes can vary dramatically from one GA application to the next, and in fact, have themselves undergone somewhat of an evolution. In early GA applications, most coding schemes were designed to produce chromosomes that were bit-strings consisting of concatenated binary sub-strings designed to represent the individual parameters necessary to form a solution to a particular problem. More recently, researchers have moved toward coding schemes which represent the various solution parameters with floating point chromosomes, resulting in chromosomes that bear a striking resemblance to arrays common to most computer languages. For many GA novices, it is this coding of the problem into chromosomes that represents the most difficult conceptual hurdle. However, once a potential GA user has been exposed to some GA applications, it is not long until said user is able to develop imaginative and effective coding schemes for a particular problem. The coding scheme plays a key role in the ultimate success or failure of a GA application.

The potential solution represented by each chromosome in the population of candidate solutions is evaluated according to a fitness function which is synonymous with the objective function of traditional optimization: a function that quantifies the quality of a potential solution. This fitness function is used in the implicit identification of high-quality values of the individual solution parameters, and the goal of the GA is to either maximize or minimize the fitness function of the strings in a generation depending on the specific problem. The fitness function ultimately determines which chromosomes are selected to propagate their parameter values through subsequent generations. Like the coding scheme, the fitness function plays a key role in the GA's success (or failure) in any given problem.

New populations of candidate solutions are generated by implementing operators inspired by natural genetic variation. The three most popular operators used in almost all GAs, are: 1) *selection*; 2) *recombination* (often termed *crossover*); 3) *mutation*.

Selection is a process through which high quality candidate solutions are chosen to form a basis for subsequent generations of solutions. Selection operators are generally driven by probabilistic decisions that ensure the best solutions are given the greatest consideration. The key idea is to select the better individuals of the population, as in tournament selection, where the participants compete with each other to remain in the population. The most commonly used strategy to select pairs of individuals is the method of roulette-wheel selection, in which every string is assigned a slot in a simulated wheel sized in proportion to the string's relative fitness. Recombination is an operation by which the attributes of two quality solutions are combined to form a new, often better solution. This operator plays a key role in determining how the quality attributes of the candidate solutions are combined. Mutation is an operation that provides a random element to the search. It allows for various attributes of the candidate solutions to be occasionally altered. Mutation takes place with a certain probability, which, in accordance with its biological equivalent, typically occurs with a very low probability. The choice of crossover and mutation probability can be a complex, nonlinear optimization problem. The mutation operator enhances the ability of the GA to find a near optimal solution to a given problem by maintaining a sufficient level of genetic variety in the population, which is needed to make sure that the entire solution space is used in the search for the best solution. In a sense, it serves as an insurance policy; it helps to prevent the loss of genetic material.

Together, these three operators produce an efficient search mechanism that generally converges rapidly to near-optimal solutions. The cycle of evolution is repeated until a desired termination criterion is reached. This criterion can also be set by the number of evolution cycles, the amount of variation of individuals between different generations, or a predefined value of fitness.

Given a coding scheme, a fitness function, and specific genetic operators, it is rather straightforward to develop a GA that mimics natural evolution to effectively drive toward near-optimal solutions. Despite the fact that the particulars of any GA application may vary, the basic approach is the same. The fundamental algorithm is summarized as follows:

1. Generate a random initial population of candidate solutions in the form of chromosomes.

2. Evaluate each chromosome in the population according to a pre-defined fitness function.

3. Employ a selection operator to create new chromosomes. The selection operator biases the new generation of chromosomes toward higher quality solutions. As the chromosomes mate, genetic operators, such as recombination and mutation, are applied to form new candidate solutions.

4. Delete members of the existing population to make room for the new candidates.

5. Evaluate the new chromosomes and insert them into the population.

6. If a satisfactory solution has been achieved (or if some other stopping criterion has been met), stop; otherwise, go to step 3.

or as programming structure:

Genetic Algorithm

```
{
*** initial time
t:=0;
*** initialize a random population of individuals
initpopulation P(t);
*** evaluate fitness of all individuals in population
evaluate P(t);
*** test for termination criterion
while not done do
                *** increase the time counter
                t:=t+1;
                *** select a sub-population of offspring
                P':=selectparents P(t);
                *** recombine the genes of selected parents
                recombine P'(t);
                *** mutation of each offspring
                mutate P'(t);
                *** evaluate the new fitness
                evaluate P'(t);
                *** select the survivors from actual fitness
                P:= survive P,P'(t);
od
}
```

Although GAs have been used to solve a wide variety of problems, most GA applications fall into one of three categories: (1) optimization, (2) machine learning in the form of learning classifier systems, and (3) genetic programming. Genetic algorithms are most appropriate for optimization type problems, and have been applied successfully in a number of automation applications including job shop scheduling, PID control loops, and the automated design of fuzzy logic controllers and NNs. The optimization applications are characterized by evaluation functions that are indicative of some desired result. Classic examples of optimization problems are those in which some profit measure must be maximized or some error term must be minimized. However, GAs have expanded their horizons well beyond traditional optimization. A second use of GAs is in learning classifier systems which employ a GA for developing effective if-then rules for solving a particular machine learning problem. GAs are used in these systems as discovery and adaptive engines that drive a rule base toward an effective solution. A third use of GAs is genetic programming, an approach to problem solving in which a GA is used to discover entire computer programs. Candidate programs are represented as tree-like symbolic expressions (S-expressions), and a GA breeds existing programs to develop new, more effective programs. As always, the search for more efficient computer programs is driven by a fitness function.

1.1   Synthesis of GA - example from robotics

As an interesting problem and example of GA application in robotics we will present the determination of topology of the neural classifier in force interaction control for robot machining operations [171], [163]. It is necessary to determine the number of neurons in each hidden layer of the neural classifier. This neural classifier represents the four-layer perceptron where its inputs are data from force sensors while outputs are the values that determine the model profile and environment parameters. In order to avoid heuristic selection of the number of neurons based on long-term simulation experiments, a new approach to network topology selection based on GAs is proposed. A first step in the application of GAs is to set the generation of an initial population of possible network topologies in a random way. In this case, it is a previously determined number of pairs, which define the number of neurons in the first and the second hidden layers. For the second step, it is necessary to convert the numeric values of the number of neurons in the hidden layers to a binary representation (two 8-bit strings). The crucial point in GA is the choice of fitness function. Our aim is to choose a topology of the neural network with the minimum approximation error, i.e., we can use the value of the well-known mean square error criterion at the end of the previously defined learning epoch as quality information for the search:

$$E(k) = \sum_{p \epsilon P} E^p(k) = 0.5 \sum_{i=1}^{k} \mid \hat{y}^p(k) - y^p(k) \mid^2 \qquad (4.1)$$

where $\hat{y}^p(k)$ is the target output of the neural network in the learning epoch k; $y^p(k)$ is the real value of network output in the learning epoch k; $E^p(k)$ is the value of the mean square criterion (fitness function for one input-output pattern p ($p \in P$) in the learning epoch k); P is the set of input-output pairs. In order to maximize the system performance, using a previously defined mean square criterion, the following fitness function is chosen:

$$ff = c - E(k) \qquad (4.2)$$

where c is an appropriate constant number chosen according to the nature of the searching task. Now, after the neural network training, all strings in the initial population have their own fitness function. Hence, according to the basic idea of "survival of the fittest", the *selection* of the genetic operator is applied. There are many selection procedures, but in this case the roulette - wheel selection [97] that selects individuals for reproduction according to their fitness function values is chosen. In view of the experience gained in the training of multilayer perceptrons, the selection procedure involves a limitation, i.e., only the pairs of strings where the number of neurons in the first hidden layer is greater than the number of neurons in the second hidden layer are ready for reproduction purposes. In order to improve the search process, the following two genetic operators (*crossover* and *mutation*) are applied with some limitations. *Uniform crossover*, which swaps each column in chromosome representation having the same probability, is chosen. In order to avoid great changes in numerical representation of the proposed problem and the proper nature of the search problem, the second operator *mutation* is limited to only five lower bits of each string. Now the complete new population is generated, which is converted into numerical representations after the decoding process, and which is ready for evaluation of its fitness function through the neural network training process with a new network topology. The process is stopped when the desired value of fitness function is attained.

For this example, the initial population of 8 pairs of the possible topology solutions is given and 3 successive generations are simulated. The following genetic parameters are chosen: crossover probability $pcros = 0.3$ and mutation probability $pmut = 0.03$. As an example, the whole evaluation process is shown in Table 4.1. Using this procedure, the following optimal network topology is selected: 6-50-20-1 (50 neurons in the first hidden layer, 20 neurons in the second hidden layer). The results in the table show the betterment process of fitness function, i.e., the convergence to optimal solution for the number of neurons in the hidden layers. Using the proposed approach and choosing the optimal node size in the hidden layers of the network, it is possible to ensure a fast learning process and better classification properties of the neural classifier.

2. GAs in Robotics

GAs can be efficiently applied in various research areas in mobile, industrial and locomotion robotics [237],[244]. The main applications of GAs are in the kinematic domain for trajectory optimization, path planning, and navigation in mobile robotics [57],[245],[384], [408], [372], [48], [119], [280], [376]. The task of navigation systems is to achieve the optimal path to an desired destination in a space with obstacles and other constraints. Traditional path planners cannot accommodate a variety of optimization criteria or allow for the changes in the optimization criteria without changing the characteristics of the planner or the search map. The genetic approach, on the other hand, can handle a variety of optimization goals such as shortest path, path smoothness and obstacle proximity, and is very tolerant to the form of the evaluation function. The GA-based path planning approaches are flexible to changes in the environment and are robust to uncertainties.

Table 4.1.   GA evaluation process for the topology of the neural classifier

INIT.POPUL.				
N0.	PAIR	CHROM.-PAR.	F.FUN.-PAR.	DIST.
1	12-15	0000110000001111	0.412	0.112
2	18-17	0001001000010001	0.403	0.110
3	16-20	0001000000010100	0.484	0.132
4	26-28	0001101000011100	0.460	0.126
5	25-16	0001100100010000	0.480	0.131
6	13-11	0000110100001011	0.497	0.136
7	12-24	0000110000011000	0.439	0.120
8	28-28	0000111000011100	0.478	0.130
MinFF=0.403	MaxFF=0.497	AveFF=0.456	SumFF=3.655	
GENER.N0.1				
N0.	PAIR	CHROM.-CHI.	F.FUN.-CHI.	
1	56-28	0011100000011100	0.489	
2	16-20	0001000000010100	0.484	
3	29-16	0001110100010000	0.384	
4	25-16	0001100100010000	0.480	
5	13-11	0000110100001011	0.497	
6	13-11	0000110100001011	0.497	
7	25-16	0001100100010000	0.480	
8	28-28	0001110000011100	0.478	
EXP.SELECT. SELECTION	0.901 0.883	1.059 1.008 43856648	1.050 1.089	0.960 1.046
N0.OF CROSS.	15		NO.OF MUTAT.	2
MinFF=0.384	MaxFF=0.497	AveFF=0.473	SumFF=3.791	
GENER.NO.2				
N0.	PAIR	CHROM.-CHI.	F.FUN.-CHI.	
1	50-20	0011001000010100	0.497	
2	16-20	0001000000010100	0.484	
3	25 -8	0001100100001000	0.456	
4	13-11	0000110100001011	0.497	
5	13-11	0000110100001011	0.497	
6	13-11	0000110100001011	0.497	
7	16-20	0001000000010100	0.484	
8	16-20	0001000000010100	0.484	
EXP.SELECT. SELECTION	1.032 1.021	0.810 1.012 21756627	1.050 1.050	1.012 1.008
N0.OF CROSS.	15		NO.OF MUTAT.	2
MinFF=0.456	MaxFF=0.497	AveFF=0.487	SumFF=3.900	

The main advantage of GAs is in great reduction of time needed for path planning. The most recent evolutionary approaches have focused on two-dimensional mobile robot path planning problems. In [57] the application of GA for the synthesis of trajectories for redundant robots, with minimization of position error in external coordinates, was demonstrated. This problem was solved using kinematic analysis, without considering robot dynamics. It is necessary to attain the path of the robot end-effector as a finite sequence of different robot joint angles which is coded in the array format:

$$[\theta_{11}, \theta_{21}, ..., \theta_{n1}, \theta_{12}, \theta_{22}, ..., \theta_{n2}, ..., \theta_{1m}, \theta_{2m}, ..., \theta_{nm}, swpos] \quad (4.3)$$

where swpos denotes the position of acceleration interrupt and beginning of deceleration; θ_{ji} is the j-th interposition of the i-th link; n is the number of interposition; m is the number of joints. For every joint, a movement is represented by an $n \times n$ grid. The movement steps are generated using a relative transition scheme, whereby from any interposition only the movement in 6 neighbouring interpositions that have great possibility of movement towards the final point of trajectory is permitted. To create the initial trajectory, one trajectory is generated in a randomized way using a relative transition scheme from the initial point, while the second trajectory is generated beginning from the final point. If these two trajectories have an intersection point, the combination of these two trajectories creates the valid initial trajectory. In the process of reproduction, the number of appearances of the same trajectory selected for crossover is limited, with the aim of greater interaction between different trajectories. Only a limited number of copies of the same trajectory that remain in the population after reproduction is permitted, while the other copies are replaced with new trajectories. The one-point crossover is chosen, where the recombination is achieved only if the place of crossover for the second trajectory is close to the circle with the centre of the place of crossover for the first trajectory. The mutation that generates the new trajectory in a randomized way is applied with the special mutation operator that changes the position of the node on the path for a small part. Greater searching space is achieved using such operators. For the fitness function, many different optimization criteria can be used such as minimal trajectory movement time, constraints on the values of joint torques, and constraints on the final joint velocities.

In the case of multi-criteria optimization based on minimal trajectory movement time, minimal energy consumption, and the actuator and environment constraints, the following combined criterion is adopted:

$$J = \sum_{j=1}^{n} h_j + \lambda_\tau \sum_{j=1}^{n} \tau_j + \lambda_v \sum_{i=1}^{6} |v_n^i| \quad (4.4)$$

where h_j is the time interval for the j-th trajectory part, τ_j is the driving torque for the j-th trajectory part; λ_j, λ_τ are the weighting factors; v_n^i the final velocity of the i-th robot joint. A special method of dynamic scaling is applied to determine the time interval and driving torques of a particular trajectory part based on the actuator constraints. The GA determines the place on the trajectory (using gene *swpos*) from which, because of the energy constraints, the deceleration process starts.

Wang and Zalzala [372] optimized the robot trajectories by GA for the industrial robot PUMA. The proposed procedure enables substantial reduction of the time needed for trajectory planning (1/12 of the time necessary when using complex search algorithms for robot trajectory planning [287]).

The first example of using GA for path planning in robotics did not contain the domain of specific knowledge. Michalewicz [245],[384] adopted the existing evolutionary planners to incorporate specific knowledge from the domain of working task. The proposed adaptive planner is capable of solving the problems of change in the environment, the problems of change in optimization criteria and requirements of the working task. The navigation is not limited only to the frame of a fixed abstract searching map of optimal path. A chromosome in evolutionary navigator (EN) is an ordered list of path nodes. Each of the path nodes, apart from the pointer to the next node, consists of the x and y coordinates of an intermediate knot point along the path, and a Boolean variable b, indicating whether the given node is feasible or not. EN unifies the off-line and on-line planning with a simple map of high-fidelity and efficient planning algorithms. The off-line planner searches for the optimal global path from the start to the desired destination, whereas the on-line planner is responsible for handling possible collisions by replacing a part of the original global path by the optimal subtour. Some special operators, based on heuristic knowledge, about the working task, were applied in order to avoid "sharp" parts of the path. Simultaneously, the tuning of operator probability is achieved using the analysis of special index of performances of these operators. Chen and Zalzala [48] applied GA for searching near-optimal paths for a mobile robot using distance-safety criteria, and they solved the multi-criteria optimization problem. They used a grid by cell decomposition to represent the environment and two numerical grid potential fields for the goal and obstacles.

For many robot controllers, there are no currently existing systematic approaches for choosing the controller parameters in order to obtain a desired performance. The controller parameters are usually determined by trial-and-error, through simulations and experimental test. Hence, in such cases, GAs were introduced as an alternative to the hand design of robot controllers, especially for autonomous robots acting in uncertain and noisy domains. The paradigm of genetic algorithms appears to offer an effective way for automatically and efficiently searching for a set of controller parameters, yielding better performance. GAs are used to search spaces of controllers described by a set of variables encoded on the artificial genotype. The fitness function is usually task-based, i.e., high scores are achieved by controllers that enable the robot to perform the desired task well.

An interesting approach to GA optimization in robotics is the tuning of control parameters for some specific robotic applications [92], [163]. The effectiveness of this approach is demonstrated by applying a simple and efficient decimal GA optimization procedure for tuning and optimization controller gains for position/force control and control of flexible link robots [92]. A robust controller based on the stability theory and using special fitness functions is developed. It is a special GA with decimal real number type representation (instead of binary type). Special fitness functions were used: integral time-multiplied absolute value of error ITAE and the normally used integral of squared errors ISE with modified versions of ITAE and ISE. Modified fitness functions were intended to penalize the overshoot and oscillation in time response of the system. The reproduction mechanism with reduction of population by 50% in each subsequent generation was adopted. Crossover was realized using randomly chosen pairs of parents in a decimal representation using the following relation:

$$k_p = r * k_{p1} + (1 - r) * k_{p2} \tag{4.5}$$

where k_{p1}, k_{p2} are the controller gains - parents of preceding generation; k_p is the controller gain - descendant; r is a random number. This genetic operator is called *operator of weighting average*. Mutation operator is applied on the selected gain using mutation probability in the following way:

$$k_p = k_p + (r - 0.5) * 2 * k_p^{max} \tag{4.6}$$

where k_p^{max} is the maximal value of the given control gain. It is clear that control parameters in GA are limited in order to prevent extremely large values of control signal because of the application of operator mutation. The evolution process is terminated when the fitness function has no great changes in the last 50 populations. In simulation experiments, important progress of the genetic process necessary for satisfying performance of the system is observed. Best results are obtained using optimization based on ITAE and a modified version of this criterion.

One of the important applications of GA in robotics is represented by the problem of object grasping with a multi-finger robot hand [72],[74],[87]. Recently, the focus has been on the effectiveness of regrasping, which enables the hand to change the contact point in grasping. The problem of generation of the grasping point and forces automatically, as the transition of the state of the target object, can be solved by GA in an effective way. The individual in genetic material is made up from the contact points on the object. The GA determines not only when and where the fingers are to move, but also which finger is operating at the next contact point with respect to the object, without losing the grasping stability. Moreover, it can generate the trajectories for each finger and the finger tip forces.

There have been many succesful applications of GA in mobile and locomotion robotics to date, ranging from simple reactive bahaviours, in wheeled robots with infrared proximity sensors [150], through visually guided behaviours in simple wheeled robots [81], to fairly complex nonreactive behaviours in simple wheeled robots [263] and a variety of locomotion controllers for six, eight or more DOF-legged robots [89], [98], [192]. [144].

It is considered that GA can be efficiently applied for trajectory generation of the biped natural motion on the basis of energy optimisation [9], [41], as well as for walking control of biped robots [49] and for generation of behaviour-based control of these systems [274].

The hierarchical trajectory generation [9] method consists of two layers, one is the GA level which minimises the total energy of all actuators and the other is the evolutionary programming (EP) layer which optimises the interpolated configuration of biped locomotion robots. The trajectory of the biped is generated using ZMP stability conditions. The chromosome on the EP level represents the interpolated configuration expressed by 12 state variables (angles) of the biped. A chromosome on the GA level consists of two parts, the first of them representing the set of interpolated configurations, while the second part includes a bit which represents the effectiveness of the configuration (0 or 1). The process runs in a cyclic procedure through the application of mutation and selection at the EP level, transfer of generated interpolated configuration into the GA level, and complete evolution process through crossover, mutation, evaluation and selection at the GA level. The fitness function on the GA level is connected to the optimization of total robot energy in order to ensure the natural movement of the biped. The fitness function also contains some constraints related to the robot motion. The final result represents an optimised trajectory similar to natural human walking, which was demonstrated by the simulation experiment.

A typical example of the application of GA in humanoid robotics was presented in [41], where the main intention was optimal gait synthesis for biped robots. The proposed method can easily be applied to other tasks like overcoming obstacles, going down stairs, etc. In solving these optimization tasks, the most important constraint included is the stability, which is verified through the ZMP concept. To ensure a stable motion, the jumping of the ZMP is realized by accelerating the body link. GA makes easy handling of the constraints by using the penalty function vector, which transforms a constrained problem into an unconstrained one. The optimization process is based on considering two different cost functions: minimization of consumed energy (CE) and minimization of torque change (TC). In this optimization process, some constraints are included, such as the stability conditions defined by ZMP to be within the sole length. The cost function based on consumed energy (CE) is defined by the following equations:

$$J = 0.5(\int_0^{t_f} \tau^T \tau dt + \Delta\tau_{jump}^2 \delta t + \int_0^{\tau_f} C dt)$$ (4.7)

where t_f is the step time, τ is the torque vector, $\Delta\tau_{jump}$ and Δt are the additional torque applied to the body link to cause the ZMP to jump and its duration time, and C is the constraint function.

In the case of minimization of the rate of change of the torque (TC), the cost function has the following function:

$$J = 0.5(\int_0^{t_f} (\frac{d\tau}{dt})^T (\frac{d\tau}{dt}) dt + \int_0^{\tau_f} C dt)$$ (4.8)

The block diagram of the GA optimisation method is presented in Figure 4.1.

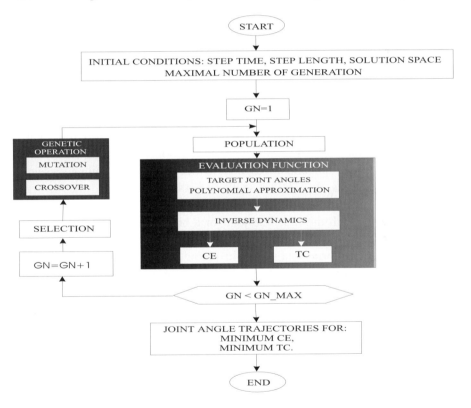

Figure 4.1. Block diagram of the GA optimization process

Based on the initial conditions, the initial population, represented by the angle trajectory in the form of a polynomial of time, is created. Its range is determined on the basis of the number of angle trajectory constraints and the coefficients are calculated to satisfy these constraints. In the simulation experiments, the parameters of real humanoid robot "Bonten-Maru I" are used. For the optimization of the cost function, a real-value GA was employed in conjunction with the selection, mutation and crossover operators. GA converges within 40 generations, while the maximum number of generations is used as the termination function. Based on simulation, the biped robot posture is straighter, like human walking, when the CE is used as a cost function. The torques change more smoothly when minimum TC is used as a cost function.

However, for the real-time applications, the process of GA optimization is time-consuming (in this case, the optimization process needed 10 minutes). Hence, the author considered teaching a RBFNN (Radial Basis Function Neural Networks) based on GA data. When the biped robot was to walk with a determined velocity and step length, the RBFNN input variable was the step length and step time, while the output variables of the RBFNN were the same as the variables generated by GA. Simulation showed good results generated by RBFNN in a very short time (only 50 ms).

In addition to biped humanoid robots, GA can be efficiently applied for transition of locomotion modes and motion planning of special snake-like robots [160]. Intelligent, biologically inspired mobile robots, and, in particular, snake-like robots have turned out to be a widely used robot type, aiming at providing effective, immediate, and reliable response to much strategic planning for search and rescue operations. In [160], a snake-like robot is described that does not have any fixed point to the environment on its body. The system is regarded as a manipulator with the first free joint or a biped locomotion robot whose ankle joint is passive. The snake-like robot has many locomotion modes to transform the shape corresponding to the locomotion mode from one shape to another without losing structural stability, whereby proper planning methodology is essential. Thus, the previous planning strategies for the manipulators based on GA cannot apply directly to the planning of the shape transition of the snake-like robot. Hence, the transformation from the initial configuration to the final configuration is divided into k intermediate configurations. The genetic search algorithm is used to find the optimal (according to the desired fitness function) set of those k configuration sequences through which the robot shape is to be transformed. Each configuration in the genotype is described as a sequence of relative joint angles of the robot body. The fitness function is defined by stability margin and smoothness properties of the particular configurations. In simulation experiments, the transition of the snake-like robot from a horizontally extended configuration to a bridge configuration like biped locomotion is considered. In GA, the elite strategy was used that implies that maximum fitness value converges around the 1500-th generation.

Another example is the application of GA to PD local gain tuning and the determination of nominal trajectory for dynamic biped walking [49]. A biped with five links is considered. In the proposed GA, 19 controller gains and 24 final points for determination of nominal trajectory are taken into account. In order that the biped body be in the vertical plane during walking, some constraints related to the fixation of joint angles are realised. Hence, it is possible to reduce the number of parameters of nominal trajectory for optimization by six. Designs to attain different goals, such as the capability of walking on an inclined surface, walking at high speed, or walking with specified step size, have been evolved with the use of GA. The fitness functions are related to total time of effective walking, average speed of the biped body, and the size of the walking step. The total number of generations for problem solving was between 10 and 60. The study showed excellent simulation results in the evaluation of control parameters, as well as in optimisation of the mechanical design of the biped.

To date, GA techniques have not been applied to a task as dynamically unstable as controlling biped locomotion. It is this inherent instability (generally two-legged walkers will fall over without continuous active control) that provides a great challenge to the hand design of such controllers, especially if smooth natural walking is required. However, given the success of evolved locomotion controllers for relatively stable hexapod and octopod robots, it was deemed appropriate to investigate the use of such techniques for developing bipedal locomotion controllers. Evolutionary robotics methods were indeed successful in finding stable controllers for biped walking. One of the main problems of GA application in humanoid robotics is coping with the reduction of the GA optimization process in real time.

One of the current applications of GA is in designing robot mechanisms [53], where the evolutionary approach is used for mass balancing, together with weighting min-max multi-criteria optimization. The aim of this approach is to find the appropriate mass and distance on the robot mechanism where compensation mass is to be placed in order to minimize the dynamic characteristics of the system (driving torques at the joints and reaction forces). Another very important and interesting application of GA is self-reconfigurable robotics [222], where evolutionary computation is used for the design of robots. Given the task of locomotion, various robots with different mechanics and control are evolved automatically.

It is important to notice that another research direction in this area is connected with a global level of optimization -*evolutionary robotics* [77]. Evolutionary robotics represents a methodology for developing robotic systems that adapt to partially unknown and dynamic environments with minimal or with no human intervention. The method is based on the artificial evolution [122] of a population of robots whose components (control architecture, electronic circuit, or morphology) are encoded on artificial chromosomes (bit strings). The best chromosomes, i.e., those that correspond to the best performing robots in the population, are selected for reproduction, crossed over and randomly mutated to generate a new population of chromosomes. The procedure is repeated until an individual with the desired chracteristics is born or until the performance of the population stops increasing [264].

Although in the early days artifcial evolution was mainly seen as a strategy to develop more complex robot controllers, nowadays the field has become much more sophisticated and diversifed. We can identify at least three approaches to Evolutionary Robotics: Automated Engineering, Artificial Life, and Synthetic Biology/Psychology. These three approaches largely overlap with each other, but still have quite different goals that eventually show up in the results obtained.

Automated Engineering refers to the application of artificial evolution for automatically developing algorithms and machines displaying complex abilities that are hard to program with conventional techniques. Within this context the desired architectures are well defined and the problem is usually cast in terms of parameter optimization by evolutionary techniques. Artificial evolution can come up with strikingly efficient and surprising solutions that exploit invariants and features invisible to an external observer.

Artificial Life is concerned with the evolution of artificial creatures that display life-like properties. In this context, the notion of evolutionary goal is not appropriate (living creatures do not evolve towards a prespecified goal) and there are very few constraints that limit the directions that evolution might take. These evolutionary systems are usually self-sufficient, self-contained, and autonomous. The selection criterion is often the energy level of the creature and the environment has an ecological validity in that it includes food sources, mates, predators, a nest, etc. These artificial worlds are easier to implement in computer simulations because simulations give more freedom to experiment with life-as-it-could-be. In this context, even evolutionary experiments that end up in the complete extinction of one species, or display alternating dynamics, such as competitive co-evolutionary scenarios [52]. [80], [307], may be considered important because they reveal interesting patterns of life. The artificial life approach is more interested in the emergent phenomena than in the optimization of a predefined strategy.

Synthetic Biology/Psychology attempts to understand the functioning of biological and psychological mechanisms by evolving those mechanisms in a robot put in conditions similar to those of the animals under study. Braitenberg [36] showed that apparently complex behaviours and emotions can be reproduced in the eye of an external observer by simple sensory-motor machines. Braitenberg was also the first person to suggest the artificial evolution of robots.

From an algorithmic perspective, an evolving robot spends most of its time interacting with the environment. Genetic operators and operations that map sensor data into motor commands may take less than 5% of the total time.

The remaining 95% is taken by mechanical actions, such as move a leg, rotate the camera, update the visual field, transmit signals across various parts of the hardware (and/or to an external workstation), etc.

The first aspect of Evolutionary Robotics concerns the level on which artificial evolution operates. It can be applied to simulated organisms or to physical robots, or to a combination of both. There is an important ongoing discussion about these issues and several strategies have been suggested to allow transfers across levels ([148], [246]) that deserve further efforts. Similarly, one can decide to evolve the control system or some characteristics of the robot body (morphology, sensors, etc.), or co-evolve them both [226]. Finally, one may decide to physically evolve the hardware, such as the electronic circuits [342] and the body shape [276]. All these issues, among others, are likely to define a new engineering methodology.

Another aspect of Evolutionary Robotics concerns the evolutionary mode. Should one use a single robot and serially test each individual one at a time [78], or is it better to use a population of such robots sharing the same environment [379]? What are the emerging dynamics and how do they affect the results? Within a collective system, one may set up a competitive scenario or a cooperative one. It may even happen that competition and cooperation autonomously develop as emergent phenomenon. Interactive mode is the situation where a robot instead evolves interactively with a human who manually selects the best individuals. There are only sporadic studies of interactive evolution, but this is going to be a crucial issue for applications related to human assistance and entertainment. Incremental mode is when one attempts to carry on evolution from previously evolved populations, usually introducing some type of modification to make the system more complex. Incremental evolution is important to tackle complex problems that cannot be evolved from scratch (the bootstrap problem), but only few studies have been dedicated to this topic so far [76], [112], [79].

An interesting approach from this area is represented by the paper of [306] where the authors implement a new strategy for improving the performance of an embedded evolutionary system developed for the automatic design of robotic controllers. An attempt is made to apply predation strategy for the first time to a population of six real autonomous mobile robots in a different way: in analogy to nature, the robot population can suffer regular attacks of a "predator" that selects the worst ("weakest") robot in the specified generation and destroys ("kills") it, opening space in the population for the migration of new individuals, hence bringing more genetic diversity to the group. This work achieves this by selecting and substituting, after a specific number of generations, the robot with the lowest fitness by a robot with a random configuration (random chromosome). Predation is a powerful strategy to prevent the population from becoming stuck in a local optimum, since it introduces new genetic material that may help the population to crawl down the slope and explore new possibilities in the fitness landscape. Becoming stuck in local optimum is an intrinsic problem of most real evolutionary systems, once it is very difficult, and some times impossible, to know from the point of view of the evolving individuals if the population can be improved even more, or if the optimal solution has actually been achieved. With the developed predation strategy, the population can count on a steady supply of new genetic material to bring in more diversity, even after it completely converged to a local optimum. It is able to achieve obstacle avoidance behaviour with the robot population evolving while deployed in the field, instead of just using the evolving group to develop an optimum controller for a single robot. This evolutionary system innovates for it can produce not only a trained robot but also an open-ended evolution, continuously adapting the robot controllers to cope with a variable environment.

The predation technique may vary according to different approaches to select, destroy, and replace the individuals of a population. For example, only one individual may be selected to be destroyed every time the virtual predator attacks, or the attack may destroy a group of individuals. All the individuals of the selected group may be destroyed, or just a smaller random number of them. The destroyed individuals may be replaced by random ones, or by the offspring of the selected breeding parents. The frequency of the attacks is another important factor, for enough generations must be left undisturbed to allow the population to recover from the attacks.

A RAM neural network, consisting of an n-tuple classifier with 28 x 2-input neurons, implements the evolvable robot control due to its good redundancy. The neuron contents are directly encoded onto a linear bit-string genotype containing 112 bits (28 neurons times 4 bits per neuron). The physical characteristics are encoded in another 26 bits (10 bits define the speed levels slow and fast and 2 bits are used to enable or disable each one of the 8 sensors). These bits are read from the chromosome in the specified order, the first 112 configure the neuronal controller (the evolving controller), the following 10 the motor control module, and the remaining 16 configure the sensor control module (the robot morphology). The robots work in a cyclic procedure, where they have a working phase, where they try to perform the selected task, and a mating phase, where they reproduce.

The results obtained with the help of the simulator were essential in providing vital insights into developing new strategies, such as predation, that considerably improved the performance of the system. The simulator also made possible rapid evaluation of different parameters, such as different mutation rates, reproduction and selection strategies.

The developed evolutionary system, helped by predation, succeeded in evolving real robots, initialised with random controllers and morphologies. The experiments demonstrated that this embedded evolutionary system was able to successfully evolve a neural network controller together with the morphology of the robots in real time in the real world. It produced a satisfactory collision-free behaviour on average after 200 generations in 60 seconds. The new genetic material it supplied in the first thirty generations was essential to allow the population to explore the fitness landscape more widely. The disadvantage of this strategy is that it never allowed the average of the population fitness to reach the maximum score, since a random robot was introduced every 10 generations, causing the performance of the population to drop. As the number and position of the sensors and the speed levels of the motors are under evolutionary control, not only the control circuit is produced, but also the physical characteristics of the robots can change into different configurations according to the complexity of the environment. In addition, the designer can fix the number of sensors, for example, and let evolution decide where they should be placed.

Most evolutionary robotics work has used evolution exclusively during the training phase of the robot controller, and at the end of the training period the controller has been downloaded to the robot and used. In [365], a combination of a training phase with lifelong evolution was proposed, removing the responsibility from the developer of either producing an accurate training phase, or initialising an adaptive robot with good behaviours. It was hypothesised that the combination of both training phase and lifelong evolution will provide a powerful way for robots to adapt and constantly re-adapt to their world.

Lifelong evolution was used to adapt this general behaviour to new environments as it encountered them. It was found [365] that in a gradually changing world the evolution strategy was successful in improving the behaviour of the robot compared to no lifelong evolution. In a catastrophically changing environment however, the evolution strategy made less improvement. Lifelong evolution was adapting to the same dynamic conditions that were present in the training phase. In the real world, more factors are likely to be dynamic than the obstacle density alone. For instance, the types of obstacles encountered, and the objective of the robot when it finds the goal could change. The real strength of lifelong evolution may lie in evolving for the unpredictable changes that occur in the world and this is a future task for investigation.

The difference of Evolutionary Robotics from other machine learning approaches is that here robots self-organize while freely interacting with their own environment. The success of Evolutionary Robotics will ultimately depend on its ability to generate new robotic systems that could not be designed with conventional techniques. This means the ability to generate robots that display complex skills and which can cope with unpredictable changes.

Chapter 5

HYBRID INTELLIGENT APPROACHES IN ROBOTICS

1.   Basic Ideas of Neuro-Fuzzy Approach

Although fuzzy logic can encode expert knowledge in a direct and easy way using rules with linguistic labels, it often takes a lot of time to design and tune the membership functions which quantitatively define these linguistic labels. Wrong membership functions can lead to poor controller performance and possible instability. An excellent solution is to apply learning techniques by neural networks, which can be used to design membership functions automatically, simultaneously reducing development time and costs and improving the system performance. These combined neuro-fuzzy networks can learn faster than neural networks. They also provide a connectionist architecture that is easy for VLSI implementation to perform the functions of a conventional fuzzy logic controller with distributed learning abilities.

Most of the proposed neuro-fuzzy networks are in fact Takagi-Sugeno controllers [330], where the consequent parts of linguistic rules are constant values. Figure 5.1 shows the commonly used connectionist fuzzy system. The system has a total of five layers. Nodes in layer one are input nodes (linguistic nodes), which represent input linguistic variables. Nodes in layer two act as membership functions (in Figure 5.1, it is Gaussian functions) to represent the terms of the respective linguistic variable. Each node in layer three is a rule node, which represents one fuzzy rule, i.e., the fuzzy operation "AND" is performed in this layer. Thus, all the nodes in the three layers form a fuzzy rule base. In layer 4, normalization based on defuzzification height method is performed. Layer five is the output layer. The first three layers define the premise of the fuzzy rule, while the links in layers four and five define the function of the connectionist inference engine.

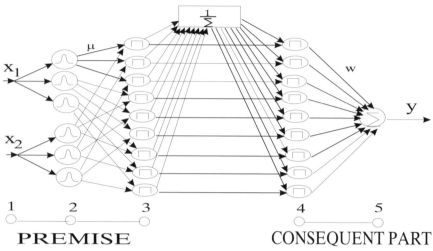

Figure 5.1.   Structure of neuro-fuzzy network

For this structure of neuro-fuzzy network, there are many different supervising learning methods [125][219], [371], [38], [39]. Many learning methods represent the applications of the back-propagation algorithm to the neuro-fuzzy system [371], [39]. Some techniques also apply GA for learning parameters of neuro-fuzzy controllers [39], [33]. Moreover, gradient-descent techniques, reinforcement learning [220],[218], and some hybrid learning techniques [151], [219],[95] have been proposed. In hybrid learning methods, the consequent part of the rules is separately tuned from the premise of the rules. The a priori knowledge about the system is used to determine the initial partition of input domains. The consequent parts of the rules are tuned by gradient-descent methods or by recursive least squares method until the desired accuracy of the output is achieved. The parameters of the premise of rule are then tuned by first-order approximation or with randomized optimization, or using GA.

One of the most important methods is Adaptive-Network-Based Fuzzy Inference System (ANFIS) [151]. The learning rule is a hybrid method which combines the gradient descent and the least squares estimate to identify the ANFIS parameters. Usually neuro-fuzzy networks are trained by applying hybrid techniques whereby the consequent parts of the rules are adapted with a supervised method and the parameters of the antecedent parts are updated with an unsupervised technique (vector quantization)[95]. The idea comes from the field of Radial Basis Functions neural networks.

Using the proposed structure of neuro-fuzzy network, the following linguistic rules of the Takagi-Sugeno controller are defined:

$RULE_1$: IF x_1 is A_{11} and x_2 is A_{12} THEN y is w_1.

...

$RULE_n$: IF x_1 is A_{n1} and x_2 is A_{n2} THEN y is w_n.

where x_1, x_2 are the inputs, y is the output of the neuro-fuzzy network, n is the number of linguistic rules, A_{nm} is the linguistic value of the n-th rule for the m-th linguistic variable, while w_n is a constant, i.e. a crisp set.

The following membership functions (Gaussian functions) are proposed:

$$\mu_{nm}(x_m) = e^{\frac{-(x_m - c_{nm})^2}{2 * s_{nm}^2}} \tag{5.1}$$

where c_{nm}, s_{nm} are the centre and the width of the membership function.

The meaning of the premise of rule using T-norm is given by applying the operator of algebraic multiplication:

$$h_1 = \mu_{11}(x_1)\mu_{12}(x_2),$$

....

$$h_n = \mu_{n1}(x_1)\mu_{n2}(x_2), \tag{5.2}$$

Applying defuzzification by height method, the following outputs are obtained:

$$y = \frac{\sum_{n=1}^{N} h_n * w_n}{\sum_{n=1}^{N} h_n} \tag{5.3}$$

The learning method based on minimization of the criterion function will be applied for this neuro-fuzzy structure. The criterion function is represented in the form of deviation of the real output from the desired output of the neuro-fuzzy network:

$$V = 0.5(y(t) - y_d(t))^2 \tag{5.4}$$

where $y_d(t)$ is the desired output of the network. In order to minimize the square criterion function it is necessary to find the gradient of the function V using parameter adaptation. For the Takagi-Sugeno controller, three basic parameters can be tuned:

1) centre of the Gaussian membership function - c_{nm}
2) width of the Gaussian membership function - s_{nm}
3) output values of rules - w_n

Tuning of the mentioned parameters is realized by the recursive gradient method:

$$c_{nm}(t+1) = c_{nm}(t) - G\frac{\partial V}{\partial c_{nm}} \tag{5.5}$$

$$s_{nm}(t+1) = s_{nm}(t) - G\frac{\partial V}{\partial s_{nm}} \tag{5.6}$$

$$w_n(t+1) = w_n(t) - G\frac{\partial V}{\partial w_n} \tag{5.7}$$

where G is the learning rate.

Partial derivatives of particular parameters are given by the following expressions:

$$\frac{\partial V}{\partial c_{nm}} = (y(t) - y_d(t))\frac{h_n}{\sum_{n=1}^{N} h_n}(w_n - y(t))\frac{x_m - c_{nm}}{s_{nm}^2} \tag{5.8}$$

$$\frac{\partial V}{\partial s_{nm}} = (y(t) - y_d(t))\frac{h_n}{\sum_{n=1}^{N} h_n}(w_n - y(t))\frac{(x_m - c_{nm})^2}{s_{nm}^3} \tag{5.9}$$

$$\frac{\partial V}{\partial w_n} = (y(t) - y_d(t))\frac{h_n}{\sum_{n=1}^{N} h_n} \tag{5.10}$$

By applying the previous equations with iterative learning procedure with presentations of the input-output training pairs and calculation of the network outputs and parameters adaptation, it is possible to tune the parameters so that the Takagi-Sugeno controller can approximate any nonlinear function with sufficient accuracy.

2.   Neuro-Fuzzy Algorithms in Robotics

Two different neuro-fuzzy approaches in robotics can be considered: first with serial connection of separated neuro and fuzzy controller, where one technique generates the necessary input characteristics for the second intelligent technique, and second approach, using the integrated Takagi-Sugeno controller for realizing collective control, the reasoning and training process.

One of the most important applications of the first group of neuro-fuzzy networks in robotics is in the field of mobile robotics [215],[346], [344]. A mobile robot is a non-linear plant which is difficult to model. The state variables of a mobile robot are easy to visualize: they have an intuitive relation to the robot behaviour. Therefore, the linguistic if-then rules can be defined in an intuitive way. The problem occurs when a robot has many sensors and actuators. The complexity of the controller increases and the construction of the rule base is more complicated, especially if a complex behaviour is required. Hence, neuro-fuzzy networks are specially applied for the complex task of mobile robot navigation and obstacle avoidance in real time. The input data to the network are based on direct or indirect data from many laser, infrared and ultrasonic sensors and some other robot velocity sensors, which measure the robot distance from the obstacles in the environment, heading angle between the robot and specified target, and velocity of the robot. The output values of the network are control signals for the robot wheels in order to determine the appropriate direction of motion and velocity of the robot. The learning of appropriate behaviour in the training process defined by the neuro-fuzzy controller can be accomplished by the supervisor or using reinforcement learning. Thus, by learning, the designer can extract fuzzy rules.

Behaviour-based control shows great potential for robot navigation in an unknown environment because it is not necessary to build exact models of the working environment and there is no need to have a complex reasoning process. Control strategy consists of two levels (Figure 5.2): 1) higher level for perception of the working environment and 2) lower level for behaviour-based control of the mobile robot. On the higher level, the neural network is used for processing information about the distribution of obstacles in the local regime of movement, while on the lower level the fuzzy controller is used to define every type of behaviour for the mobile robot, to solve conflicts and manage the competition between different types of behaviour.

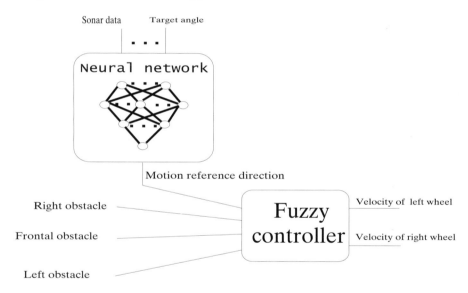

Figure 5.2. Neuro-fuzzy architecture for mobile robot navigation in an uncertain working environment

The inputs to the neural network are the target angle between robot and specified target and the data from the sonar $d_i, i = 1, .., 15$ that detects the obstacle from the left, right, and the front. The output of the neural network is the motion reference direction needed for robot navigation. The training process of the four-layer perceptron is realized by back-propagation algorithm, using test data that are similar to the data in real-time navigation. Based on the motion reference direction and distance between the obstacles and the robot in the dynamic working environment, the behaviour fusion of the mobile robot is achieved using fuzzy reasoning and control of the left and right wheel velocity. This behaviour fusion is realized using weighting of particular rules related to appropriate types of behaviour (edge following, obstacle avoidance, target steering, deceleration in the case of circular paths and narrow passages) in the process of fuzzy reasoning (min-max inference algorithm). Experimental results [215] for this type of connection of neuro-fuzzy systems show that the proposed neuro-fuzzy controller can significantly improve navigation performance of mobile robots in a complex and unknown working environment. These neuro-fuzzy networks are especially useful for robot navigation in the case of multisensor fusion and integration. In mobile robotics, a neuro-integrated fuzzy controller has been proposed [261], consisting of three basic sequential parts: 1) FMF - generator of fuzzy membership functions used for fuzzification of sensor inputs; 2) RNN - neural network of rules used for interpolation of the set of fuzzy rules; 3) ORNN - neural network used for on-line tuning of output. In this way, numerical data are no longer used as training data, because the reliable expert rules are applied. Also, the fuzzy membership function ensures the continual input representation that realizes a lower sensitivity to the parameter and control goal variations. The number of rules is reduced significantly (only five rules are necessary for a wall following operation, while for corridor passing and convoy tracking nine rules are needed). The RNN is trained through 200-300 learning epochs, while the ORNN learning process is terminated in less than 100 epochs.

In [349], reinforcement learning a neuro-fuzzy controller (driving the robot motion) is used for local navigation of a mobile robot. The learning-to-drive algorithm consists of two terms: a "critic" transforming the environmental feedback to a higher-quality heuristic signal, and an "actor", actually controlling the robot, which learns to exploit the signal provided by the critic. The output value of the controller is a weighted average of the output values of each activated fuzzy rule, and it is applied directly to the motors of the robot. The reinforcement signal is available after each iteration taking values in the interval $[-2, +2]$ where the value -2 corresponds to robot failure and the value $+2$ to robot success (reaching the target within a distance of 10 length units). To improve the behaviour of the algorithm, the membership functions were allowed to be more fuzzy during the first stages of learning, and a momentum term with a high learning rate was employed in the weight updating scheme. Also, the gradient descent method was used for parameter identification of the input variable membership functions.

Another interesting example of sequential connection of fuzzy logic and neural networks is represented by the selection of reflex force gain for a bilateral teleoperation robotic system [46] (Figure 5.3).

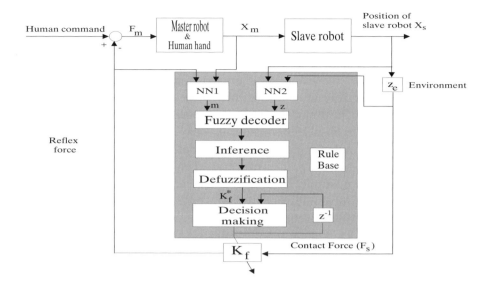

Figure 5.3. Determination of reflex force gain using fuzzy and neuro approach for bilateral teleoperation

Two neural classifiers are used to estimate characteristics of the robot hand and characteristics of the working environment with which the "slave" robot is in contact. Input data to the neural network are the data about position and force for "master" and "slave" robots. Output data of the "master" neural network represent the estimation of information m about the strength of squeeze of the robot hand using linguistic variables "SOFT" or "HARD". In the second case, output data of the "slave" neural network represent the information z about type of working environment (these data determine the working environments with different values for stiffness). Using this information, a simple fuzzy-rules base is formed, which is used for the determination of reflex force gain. This reflex force gain must change according to the characteristics of the "master" robot and working environment in order to obtain satisfactory performance of the robotic system. Because of the sensor noise and imperfect training process through the decision-making process, a change of reflex force gain for a small change of estimated characteristics is avoided. Using control of the robotic system with model approach, Huang and Lian [134] applied the hybrid neuro-fuzzy approach, which consists of a simple fuzzy controller and a coupled neural controller that is based on input data about the desired trajectory. The complete control consists of a summation of the outputs of the neural and fuzzy controller. The conventional fuzzy controller is designed for each robot joint separately and its role is to be a "rough" tuner of the system performance, while the neural controller reduces the coupling effects and its role is to be a "fine" tuner of the system performance. In this way, the fuzzy controller significantly improves the convergence speed of the neural controller.

In the second group of neuro-fuzzy approaches in robotics, one of the important applications is in the area of mobile robot control [95],[377]. The mobile robot is a nonlinear control object that is in most cases very hard to describe in detail by mathematical models. The state variables of mobile robots can be easily represented by visual information that have intuitive relation with the robot's behaviour. In this case, it is possible to define the linguistic rules in an intuitive way. However, the problem appears when the robot has many sensors and actuators. The controller complexity is increased, the design of the rule base is a complex problem, especially in the case of a complex behaviour of the mobile robot. Hence, neuro-fuzzy networks can be applied for complex navigation tasks of the mobile robot and for obstacle avoidance in real time. The input data are obtained in a direct or indirect way using laser, infrared and ultrasound and other robotic sensor systems that measure the robot distance from the obstacle, navigation angle between the robot and special target, and the robot speed. The output values of the network are control signals for the robot wheels in order to determine the appropriate direction of movement and robot speed. Learning of the appropriate behaviour in the training process by the neuro-fuzzy controller can be realized using either supervised or "reinforced" training as an unsupevised learning method.

In [95], an effective example of the application of a neuro-fuzzy controller to control miniature robots KHEPERA is considered. The basic configuration of the KHEPERA robot consists of the CPU (16MHz microcontroller) with a sensor-actuator board to control two DC drives for wheel motion that are coupled with eight infra-red distance sensors. The linguistic variables for controller inputs are *distance between the robot and obstacle in the direction left, right, straight, back*, while the output linguistic variables are *speed of left and right wheel*. The input linguistic variables have three linguistic values *Big, Average, Small*, while the output linguistic variables have seven linguistic values *Forward Fast, Forward Average, Forward Slow, Stop, Backward Slow, Backward Average, Backward Fast*. The linguistic variables are expressed via membership functions in the form of Gaussian functions. The neuro-fuzzy controller was trained by supervisor, whereby the supervised learning in the first phase is achieved by human hand. However, this method appeared to be very complex, with more than 500 learning iterations. Hence, in the second phase the supervisor is adopted in the form of a Braitenberg vehicle that represents a simple neural network that connects by inhibitory and exciting connection the distance sensor and the wheels of the mobile robot. The applied supervised method enables rule extraction from adapted parameters of the neuro-fuzzy network. There were efficient experiments with two different robot tasks: wall following and obstacle avoidance. The proposed method also includes the initial knowledge about mobile robots in the neuro-fuzzy controller.

In a similar approach [377], the tracking problem of desired asimuth and speed of mobile robot with two separately driven wheels is solved by two neuro-fuzzy networks (one for asimuth, the other for speed) of the Takagi-Sugeno type with separate reasoning, which are coupled at the output with neural networks of fixed weighting factors. The inputs to the neuro-fuzzy networks are the errors of asimuth and speed and their derivatives. Using 49 rules, the experimental results showed good performances in the tracking of square and circular trajectories.

An integrated self-organizing neuro-fuzzy controller is applied to control autonomous underwater vehicles (AUV) in order to serve as a better control alternative in comparison with PID control and conventional adaptive control [368], [370], [369]. Because of the nonlinear dynamics of an AUV and the difficulties in modelling its interaction with an unstructured and uncertain environment, a neuro-fuzzy controller as six-layer feedfoward network is applied, which is capable to self-organize and adapt to a new control situation. In [370], the proposed SANFIS architecture incorporates fuzzy basis functions that are capable of universal approximation. In addition, the parameters of SANFIS are separated into linear and nonlinear parameter sets, which are then optimized by a recursive least squares and a modified Levenberg-Marquardt algorithm, respectively. This hybrid learning algorithm together with an on-line clustering technique and rule examination provide SANFIS with the capability of self-organising and self-adapting its internal structure for learning the required control knowledge for an AUV to follow the desired trajectories. Furthermore, SANFIS is capable of adapting itself to the changing environment through an on-line learning algorithm. In [369], the utilization of a self-adaptive recurrent neuro-fuzzy control as a feedforward controller and a proportional-plus-derivative (PD) control as a feedback controller for controlling an AUV in an unstructured environment is presented. Without a priori knowledge, the recurrent neuro-fuzzy system is first trained to model the inverse dynamics of the AUV and then utilized as a feedforward controller to compute the nominal torque of the AUV along a desired trajectory. The PD feedback controller computes the error torque to minimize the system error along the desired trajectory. This error torque also provides an error signal for on-line updating of the parameters in the recurrent neuro-fuzzy control to adapt to a changing environment. A systematic self-adaptive learning algorithm, consisting of a mapping-constrained agglomerative (MCA) clustering algorithm for the structure learning and a recursive recurrent learning algorithm for the parameter learning, has been developed to construct the recurrent neurofuzzy system to model the inverse dynamics of an AUV with fast learning convergence. The main salient features of the proposed control scheme are: 1) the recurrent neuro-fuzzy system incorporates fuzzy basis functions with dynamic elements for better approximation of nonlinear dynamic functions, 2) the resulted recurrent neurofuzzy controller is capable of translating the complicated dynamic behaviour of a system into a set of simple linguistic dynamic rules in the state-space representation, and 3) a systematic self-adaptive learning algorithm, consisting of an MCA clustering algorithm for structure learning and a recursive recurrent learning algorithm for parameter learning, has been developed to provide the recurrent neuro-fuzzy system with fast learning convergence and a parsimonious system structure. Computer simulations of the proposed recurrent neuro-fuzzy control scheme and its performance comparison with an adaptive controller have been conducted to validate the effectiveness of the proposed approach.

Another example of using the integrated neuro-fuzzy approach in the form of fuzzy B-spline controller is in the application of visual learning for fine positioning of a manipulator into a grasping position [405]. In this approach, the principal component analysis is used to reduce the dimension of the input visual information for the neuro-fuzzy controller. It has been shown that this approach leads to a very robust system that is stable under variable environment conditions.

Neuro-fuzzy networks can be efficiently applied for dynamic learning control of position/force in robot machining operations [176], [177],[178], [180], [181]. For this type of operation, the problem is to determine an efficient position/force controller, together with obtaining the desired force and desired tool feedrate. It is hard to obtain these values in practice, especially for the deburring process, because a human expert can realize in an efficient way the desired force and toll feedrate according to the environment characteristics and burr height. It is also very hard to model the experts knowledge, hence it is better to extract flexible linguistic rules based on his knowledge, skill and experience. The inclusion of adaptive and intelligent control properties is necessary too because of the complexity of modelling the robot, tool, and time-varying characteristics of the working environment. Because of these facts Kiguchi and Fukuda [178] proposed an intelligent controller (Figure 5.4) which consists of two parts: intelligent planner and neuro-fuzzy controller of position and force. This intelligent method, without experimental work and previous learning process, enables the acquisition of expert knowledge for the deburring process. The intelligent planner consists of a neuro-fuzzy planner and a fuzzy estimator of characteristics. The neuro-fuzzy planner, which is tuned in real time, determines the desired tool feedrate. The fuzzy estimator of unknown characteristics of the working environment generates the desired force and input coefficients for the adaptive neuro-fuzzy force controller. The intelligent planner transmits the generated signals to the neuro-fuzzy controller of position and force. According to the principle of hybrid position/force control, two neuro-fuzzy networks exist based on Takagi-Sugeno architecture, one for position control and the other for force control. The inputs to the neuro-fuzzy position controller are position and velocity errors in external coordinates E and $\dot{E}$, while the inputs to the neuro-fuzzy force controller are the position errors in external coordinates E and robot moment in the force control direction $M0$. The training process for these neuro-fuzzy networks is realized by supervised back-propagation method, using measurement of position and force. The fuzzy estimator of characteristics based on measurement of position and force evaluates the stiffness of the working environment as a fuzzy variable and, using expert knowledge expressed through fuzzy rules, determines the desired force and desired tool feedrate. Another role of the fuzzy estimator of characteristics is to avoid great force overshoot based on modified coefficient of input variables for the fuzzy force controller. The neuro-fuzzy network is included in the neuro-fuzzy planner for determining the desired tool feedrate, using information about burr height given by the camera or laser sensor.

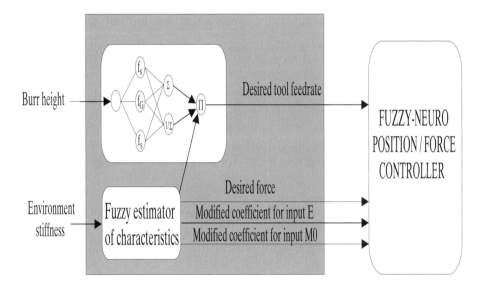

Figure 5.4. Intelligent planner for neuro-fuzzy control of robot machining operations

A further extension of the proposed approach, to deal with the unknown or unexpected environment, is given in [180]. A suitable neuro-fuzzy force controller is selected immediately from the initially prepared multiple neuro-fuzzy controllers for various kinds of environments, and then harmonized with a proper ratio using fuzzy reasoning according to the dynamic properties of the unknown environment. In this method, on-line classification of the unknown environment is carried out with an off-line trained neural network [171], then fuzzy reasoning for controller selection and harmonization is performed on the basis of the classification information. Consequently, the appropriate neuro-fuzzy controller is selected for the initially expected environment, and some suitable neuro-fuzzy controllers are selected and harmonized with a proper ratio for an unexpected environment.

Young and Shiah [393] proposed an interesting method to enlarge learning space coverage in robot learning control, based on information storage via effective memory management. In this approach, a new structure for learning control is proposed to organize information storage via effective memory management. The proposed structure is motivated by the concept of the human motor program and consists mainly of a fuzzy system and a cerebellar model articulation controller (CMAC)-type neural network. The fuzzy system is used for governing a number of sampled motions in a class of motions. To allow for automatic adjustment of the system parameters, the fuzzy system is implemented with the structure of a fuzzy-neural network. The CMAC type neural network is used to generalize the parameters of the fuzzy system, which are appropriate for the governing of the sampled motions, to deal with the whole class of motions.

A fuzzy system is used to represent the abstract motor program and a CMAC-type neural network is used to manage the parameters. The parameters specifying the fuzzy rules for governing sampled motions are stored and manipulated by the CMAC-type neural network to deal with a wide range of motions. In order to simplify the complexity in learning, robot motions were classified in advance according to their features. As an example, a group of arbitrary robot motions can be categorized into a class of motions of various movement distances with the same movement velocity and load or a class of motions of various movement velocities with the same movement distance and load, etc. Under this arrangement, a class of motions with the same feature are expected to correspond to similar fuzzy parameters. Consequently, the data with which the CMAC-type neural network (which executes the generalization) is to deal will exhibit less nonlinearity. Parameters of the fuzzy system for various classes of motions can also be incorporated into the same neural network at the expense of greater memory requirements. For instance, the neural network can store the parameters corresponding to a group of motions, which reach different destinations with different velocities and loads. Naturally, a fuzzy system is not the only choice for representing a motor program. For instance, the gains of a conventional PD or PID controller or the torques to move the robot links can also be generalized by a neural network to govern various robot motions. However, the authors believe that generalization of qualitative fuzzy rules is more effective than that of quantitative numerical data, because the former involves the generalization of abstract representations and tends to cover more learning space.

One of the reasons for adopting the CMAC-type neural network rather than some other type is that its learning for certain motions will not affect that for other motions too much. Thus, training patterns can be added or deleted easily according to the performance. In addition, this type of network has a strong generalization capability and a simple structure. Under this design, in some sense the qualitative fuzzy rules in the fuzzy system are generalized by the CMAC-type neural network and then a larger learning space can be covered. Therefore, the learning effort is dramatically reduced in dealing with a wide range of robot motions, while the learning process is performed only once.

As the proposed scheme can be used to govern a class of motions with the same feature, possible industrial applications can be the tasks that involve a number of workpieces with different loads, movement distances, etc. In addition, the scheme also provides the flexibility that workpieces in the same class can be added during task execution without repeating the learning process. Simulations emulating ball carrying under various conditions were presented to demonstrate the effectiveness of the proposed approach.

Because of their complementary capabilities hybrid intelligent methods have also found their place in research on gait synthesis and control of humanoid robots.

In [156], a learning scheme based on a neuro-fuzzy controller to generate walking gaits, is presented. The learning scheme uses a neuro-fuzzy controller combined with a linearized inverse biped model. The training algorithm is *back propagation through time*. The linearized inverse biped model provides the error signals for back propagation through the controller at control time instants. For the given prespecified constraints, such as the step length, crossing clearance and walking speed, the control scheme can generate the gait that satisfies all the mentioned constraints. Simulation results have been verified for a simple structure of five-link biped robot.

The neuro-fuzzy approach can be efficiently appplied in the area of aerial robotics. In his research, Montgomery [253] addresses the helicopter controller synthesis and tuning problem. A model-free "teaching by showing" approach is used to train a fuzzy-neural controller for autonomous robot helicopters, in simulation and hardware. A controller is generated and tuned using training data gathered while a teacher operates the helicopter. This approach is useful for time-varying systems for which mathematical models are unknown but which can be stabilized and controlled by a human operator. The methodology uses techniques from the fields of behaviour-based control, fuzzy logic, neural networks and teaching by showing, all of which are model-free. The controller is decomposed by a human expert into a hierarchical behaviour-based control architecture with each behaviour implemented as a hybrid fuzzy logic controller (FLC) and general regression neural network controller (GRNNC). The FLCs and GRNNCs are generated through teaching by showing and they share in the control task. The FLCs are built during initial controller generation, remain static once created and provide coarse control of the helicopter. The GRNNCs are incrementally built and modified whenever the controller does not meet performance criteria, they are dynamic and provide fine control, enhancing the control of the FLCs. The methodology is applied both in simulation and on a radio controlled (RC) model helicopter for real world validation. In simulation, roll and pitch controllers were generated and tuned. They were shown to be capable of meeting performance criteria for both noise and noise-free test cases. However, when tested on actual hardware the approach was inadequate. A roll controller, generated using teaching by showing, could not meet desired performance criteria. Hence, it is necessary to upgrade the proposed theoretical approach for real world solutions.

2.1   Hybrid genetic approaches in robotics

In cooperation with neural networks, GAs can be effectively used for the determination of optimal weighting factors, topological characteristics of the network (choice of the number of neurons, number of network layers, types of activation functions) and parameters of the learning algorithms. On the other hand, neural networks can be evaluation functions for GA in the case of complex optimization problems.

GA has been efficiently applied in robotic neural approaches, as in the case of the neuro-GA controller for visually-guided swing motion of a biped with 16 DOFs [258]. The aim of this robot task is learning of swing motion by neural network using visual information from a virtual working environment. As is known, GA requires a lot of computing time in order to evaluate the fitness function for each individual in a population. Hence, it is not desirable to use direct execution of the working task on a real biped because of task complexity and inaccuracies of the sensors. Instead of a real biped, a virtual working environment is used for acceleration of the learning process. As the learning process is transferred from the virtual environment to the real robot, the difference existing between these two systems is neutralized by generalisation capabilities of the neural network. The aim of learning for visually guided swing motion is to increase the swing amplitude by skillful change of the gravity centre of the biped robot in the direction of the swing radius, caused by dynamic change of the environment recognised by the vision sensor. The input to the network represents sensor information from the vision sensor, while the output of the neural network is the knee angles of the biped (Figure 5.5). GA optimises the three sets of weighting factors of this 4-layer neural network. At the output of the network, the data are transformed into joint angles and then, using limiters of angular velocities (to avoid extreme changes of joint angles), the knee joint angle is calculated. The genotype is represented by a sequence of weighting factors. The number of individuals in the initial population is 200. The fitness function is represented by the height of the centre of gravity in the initial and final pose. The evolution simulation experiment is terminated when the number of alternations in generations reaches 50. The results show the efficient learning of swing motion through successive generations that has been verified through generalisation experiments on the real robot biped.

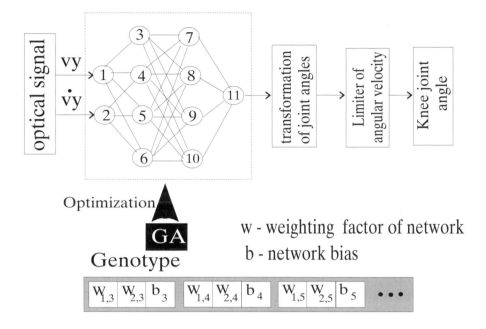

Figure 5.5. Neuro - GA approach for optimization of robot swing motion

In [86] the authors deal with a GA application for the determination of weighting factors of a recurrent neural network in order to generate a stable biped gait. When the biped robot walks on ground with some gradients, the optimal trajectory is not known, hence the optimal trajectory of ZMP is not realised. Because of this, reinforcement learning is used by applying a recurrent neural network. A recurrent neural network is chosen in order to select best biped configuration (desired joint position and velocity) using ZMP as stabilisation. This type of neural network was chosen because the output of the network generates the dynamic output data for static inputs and can describe time records easily. The input to the network is the information about position of ZMP taken from the force sensor, while the output of the network is the correction angles and correction velocities needed for a stable motion. The ZMP is calculated using the values from force sensors at each sole and values of joint angles. Only self-mutation is used from the set of genetic operators based on addition of the Gauss noise with multiplication by the value of fitness function. The elite selection is chosen, while the fitness function is defined by the sum of squares of the deviations of the desired coordinates from the ZMP coordinates. In both single-support and double-support phases of walking the algorithm calculates ZMP by using values from four force sensors at each sole, while correction to actuation angles and velocities is determined by recurrent neural networks with the ZMP being within the supporting area of the sole of the robot. A block diagram of the stabilisation biped control is shown in Figure 5.6, where θ, $\dot\theta$ are the joint angles and velocities; θ_D, $\dot\theta_D$ are the desired joint angles and velocities; U is the control signal; F is the foot force. The motion on inclined surfaces is investigated with initial population of 50 different individuals. It has been shown that the use of this approach yields a stable biped gait.

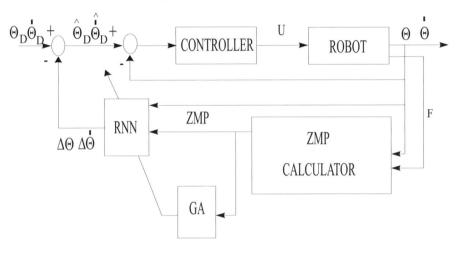

Figure 5.6.   Block diagram of stabilization biped control

Reil and Husbands [281] proposed an evolutionary approach for the biped controller based on a dynamic recurrent neural network. Each neural network consists of ten fully interconnected neurons. The first six neurons represent motor neurons because that they control biped actuators (the biped has six DOFs). Their outputs are scaled to map to the angle limits. The GA has the task to optimize the weighting factors, time constant, and bias of activation functions for the chosen neurons. Parameter values are coded using real numbers with different ranges for each type. Each population consists of 50 individuals. The Rank-Based selection is used for generating new generations with a fittest fraction. From the genetic operators, only mutation with a small rate is applied. The fitness function considers two components: 1) the minimisation of travelling distance from the origin; 2) the gravity centre cannot be below a certain height. The fraction of evolutionary runs leading to stable walkers was 10%, of which the average walking distance was 20.577 m. All controllers in the simulation experiments walked in a straight line without the use of proprioceptive inputs and without active balance control, because of the application of the mentioned fitness function. The authors proposed an additional criterion for fitness function in order to reward cycling activity. In this way, the proportion of successful runs is increased to 80% but without improvement of the overall time efficiency. In order to achieve walking on a rough terrain, sets of simulation experiments with the inclusion of sensor signals as input to neurons of the neural network, are realized. These preliminary experiments on integration indicate that cyclic walking activity can indeed be modified in a meaningful way by external stimuli. The quality of simulation results indicates a further improvement by a refined fitness function, together with inclusion of coupled neural oscillators instead of a single neural network. It is desirable also to incorporate biomechanical knowledge about human walking. However, because of the existence of only simulation results, a the implementation of the proposed theory on the embodied humanoid robots is a real problem.

A very interesting approach was proposed in [71], using the ideas of artificial life. The main idea is to optimise both the morphology and control of biped walking at the same time, instead of optimizing the walking behaviour for the given hardware. It was shown that the generated robots have diverse morphologies and control systems, while their walking is fast and efficient. Both the morphology and neural systems are represented as simple large tree structures that are optimized simultaneously. From the morphology side, the lengths of the lower and upper limbs are optimized. Two types of control systems are analyzed: one based on a neural network and the other based on a neural oscillator. The input to the neural network represents the velocity, acceleration, and ZMP position, while the output of the network represents the joint angles. It is a layered neural network with a pair of hidden layers. The chromosome includes the following parameters for optimization: information on initial angle and velocity, length of each link and weights of each neuron in the neural network. The simulation experiments with population size of 200 individuals and 600 generations were realized using standard genetic operators. In the first phase of GA, the fitness function was the distance between the centre of mass of the robot and the initial point. In the second phase, two fitness functions were evaluated based on the efficiency and stability of walking. The preferred solution has appropriate locomotion and morphology. As the other solution for control algorithm, a neural oscillator was used, because biped walking is a periodical and symmetrical solution. The neural oscillator generates the rhythm for the biped walking. In this case, it is not necessary to use a large-size GA chromosome, as was the case with neural networks. The structure of the neural oscillator represents some kind of recurrent neural network dynamic state, while the other parameters of the GA optimization process are the same as in the previous case. The walking patterns for the neural network and neural oscillator are similar, even though the sizes of the chromosomes are much different (the neural network chromosome has 1000 bits size, while the size of the neural oscillator is 300 bits). Therefore, a larger dynamic model of biped can be applied to the model with a neural oscillator. It has been shown that there is a close relation between the morphology and locomotion. The proposed co-evolution of the morphology and control systems may be a potentially powerful method for designing real humanoid robots.

Researchers at the Center for Neural Engineering at the University of Southern California [213] have used GA methods to evolve the weights for a neural controller for a robotic hexapod RODNEY. Fitness functions were defined first for learning oscillatory control for each of the legs and then coordinating the oscillations to produce effective gait.

Genetic algorithms have also found application in the area of behaviour-based robotics, to control robots in tasks requiring sequential and learning behaviour [390]. The Family Competition Evolutionary Algorithm (FCEA) is applied on a set of recurrent networks with the aim that a chosen network can integrate different types of behaviour in a smooth and continuous manner. An important characteristic of the algorithm is the integration of special selection and operator techniques such as family competition, decreasing-based Gaussian mutation, self-adaptation Gaussian mutation and Cauchy mutation. Each of the mentioned strategies in FCEA can compensate for the shortcomings of each other. In order to illustrate the power of this approach, two different special robotic tasks were investigated: the "artificial ant" and a sequential behaviour problem - an agent learns to play football. This approach outperforms the other genetic approaches for the mentioned examples. In another work from this area, a Braitenberg style neural controller [78] was implemented on a small commercially available KHEPERA robot equipped with three ambient light sensors pointed at the floor and eight infrared proximity sensors. Fitness functions were defined for various behaviours, including navigation and obstacle avoidance, homing and grasping of balls using an attached gripper.

In cooperation with fuzzy logic, GA can be efficiently applied for optimization of the shape of membership functions, set and number of fuzzy rules and scaling factors. GA can be also used for structure optimization of neuro-fuzzy networks by tuning of weighting factors, parameters of membership functions and parameters of learning methods [139]. In the application of GA for fuzzy controllers, it is necessary to solve some problems. The first is transformation (interpretation) between the domain of fuzzy knowledge and GA coded domain. The other problems include the way of using initial expert knowledge for better and faster search, as the way of using initial knowledge in the case when more experts exist. An important problem in practice is related to fast implementation of the fuzzy-GA controller in real-time control.

To solve the above problem, an approach to the synthesis of a GA adaptive fuzzy hierarchical controller has been proposed [3]. This fuzzy-GA controller uses the spatial data and data from sensors to control a flexible robot (Figure 5.7).

The GA performs optimization of two fuzzy systems: fuzzy feature extractor in the higher control layer and fuzzy controller in the lower control layer. For the first fuzzy controller, linguistic variables are defined using shape features of the robot arm: *straight, oscillatory and positive and negative gently curved*. These linguistic variables are represented by the parameters of mean and variance of curvature taken from space measurement using strain gauges. For the second fuzzy controller, using the mentioned four fuzzy linguistic variables, the linguistic variables of position and velocity tracking error taken from the robot sensors are used. In this way, there are 36 possible fuzzy rules. Only three parameters of membership function are coded for optimization. In order to include initial expert knowledge, the "grandparent" scheme is applied. All members of initial population represent only binary mutation of the "grandparent". In order to include knowledge of more experts, methods with simultaneous evolution of more different population are applied. For this problem there exists an interesting hardware solution where the fuzzy controller works on a DSP board in direct connection with the GA, which is executed on a Pentium 133 MHz board. A special fitness function is selected to represent the integral, which is inversely proportional to the positional tracking error and tracking overshoot. The results show that the rising time of the fuzzy-GA algorithm is smaller by a factor of 11 compared with conventional fuzzy algorithm.

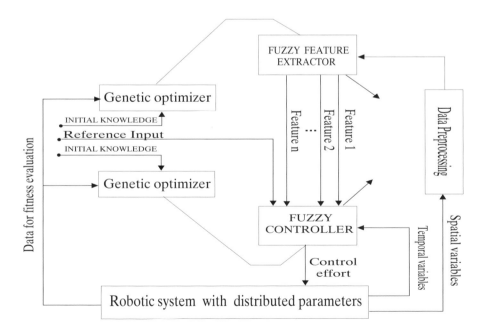

Figure 5.7.   Fuzzy-GA training hierarchical control structure

In [243], a simple and efficient application of GA for fuzzy control of manipulation robots is given. The control algorithm is represented by a fuzzy controller only, using GA for optimization of the set of fuzzy rules and parameters of membership functions. The optimized parameters are represented by a chromosome with a total of 45 bytes. Using standard genetic operators, optimization is performed based on the criterion of weighted integral of position and velocity tracking error. In a similar fashion, in [153] Jin used GA for optimization of a fuzzy feedforward computing torque system for decentralized control of robot manipulators. In this case, the improved GA was not used only for optimization of parameters (parameters of membership functions in fuzzy rule base of dynamic modelling of the robotic system), but it was also used for structure optimization of the fuzzy system (optimization of rules in the rule base).

Another typical application of the fuzzy-GA approach in robotics involves tuning of the local fuzzy-PID controller gains [123]. As is well known, the procedure of gain tuning is not systematic because in most cases gain tuning is achieved through trial-and-error. On the other hand, complex system expert knowledge is not sufficient for satisfactory performance. For this controller, fuzzy logic variables at the input are position, velocity, and integral tracking error, while the fuzzy logic variables at the output are position, velocity, and integral gain. The input linguistic variables are represented using the linguistic values *small, average and big*, while the output linguistic variables have a broader range: *very small, small, almost small, almost average, average, almost big, very big, huge, very huge*. These variables are represented by Gaussian membership functions which have two parameters for optimization: mean and standard deviation. Each GA individual represents the complete rule base with 81 rules. A specific combined fitness function is adopted, which consists of inversion of more particular fitness functions as the sum of control difference, cube root of position tracking error sum and average change of integral gain until reaching the rising time. Simulation results are based on the generation of 80 successive populations, or of a smaller number if the results are satisfactory. The fuzzy-GA controller showed good results in tracking and robustness in comparison with conventional fuzzy controller and conventional PID controller.

Go and Lee [94] showed that a self-tuning fuzzy infererence method by GA can be efficiently applied for the fuzzy-sliding mode control of a polishing robot. Usung this method, the number of inference rules and the shape of membership functions are optimized without a human expert. It is guaranteed that the selected solution is the global optimal solution based on optimization of the Akaike information criterion expressing the quality of the inference rules.

An especially interesting range of fuzzy-GA applications is in the domain of mobile robotics and autonomous vehicles [110], [385],[199]. In [199], a fuzzy-

GA controller is applied to stabilize the motion of a mobile robot. The aim of GA is to select and update the output parameters of fuzzy rules. Because of this property of GA, asymptotic stability cannot be guaranteed. Hence, some kind of supervisory control is introduced in order to restrict the derivative of the proper Lyapunov function to be negative semi-definite. The fitness function is defined by the desired Lyapunov function and derivative of Lyapunov function. The Automated Tuned Evolutionary Programming technique is used for generating offspring based on the mutation operator. The simulation results show that the tracking error decreases exponentially because the GA drives the controller toward the hypersurface where the exponential stability is guaranteed.

GA was efficiently applied to automatically generate fuzzy if-then rules for the fuzzy controller of a mobile robot in a manufacturing cell [385]. The fuzzy-GA controller is developed for intelligent path planning with the aim to automatically realize fuzzy rules from the samples, which makes the fuzzy rules base rich and completely continuous. The initial population of rules is formed from two groups: one is converted directly from the fuzzified train case, which is usually accurate but rather specific, and the other is generated at random. The genotype consists of the sequence of linguistic terms with the binary notation 0 or 1, which represent the valid and invalid linguistic terms. The roulette wheel selection is applied for reproduction purposes. The crossover and mutation sites are determined by a random number generator.

A special application of the fuzzy-GA approach is representes by the development of a fuzzy vehicle controller for real crop harvesting in agriculture [110]. The advantage of this approach is in reducing the number of rules, facilitating a better behaviour arbitration and rule learning in which reinforcement is given as actions are performed. The system uses sensory information in order to narrow the search space for the GA. If the fuzzy rule base contains poor rules (deviating from its objective), the on-line algorithm is fired to generate a new set of rules to correct this deviation. The fitness function is evaluated based on information about how much GA reduces the absolute deviation. The fuzzy-GA control architecture was evaluated in an in-door and out-door environment when the robot was able to track various edges under different environmental and ground conditions.

In [181], GA is applied in combination with neuro-fuzzy force controllers for robotic contact tasks. In this case, multiple genetic neuro-fuzzy force controllers are suitably combined with a proper rate in accordance with the unknown dynamics of an environment. In order to carry out the proposed force control method, several kinds of neuro-fuzzy controllers designed for different kinds of environment are prepared. The optimal combination rate of the prepared neuro-fuzzy force controllers according to the environment dynamics is defined on-line by a neural network that is off-line trained using genetic algorithms. Exactly, optimal weights of this Controller Combining Neural Network are defined by using GA to output proper weight values for each neuro-fuzzy force controller in accordance with the dynamic properties of the environment.

Chapter 6

SYNTHESIS OF CONNECTIONIST CONTROL
ALGORITHMS FOR ROBOT CONTACT TASKS

1.   Introduction

Control of position and force arising in the contact of the robot manipulator
and its dynamic environment is a basic requirement in many robotic tasks in
industrial practice. Strong motivation for studying this problem is gained from
new, complex and somewhat unspecified robotic tasks such as assembly, machining, and remote handling in an unstructured environment, where automation
currently represents one of the most challenging topics in robotics. A common
control problem in this type of tasks is how to describe the robot-environment
dynamics in a correct way and synthesize control laws, which would stabilize
both the desired position and the interactive force. As is well known, many
manipulation robots have to operate in uncertain and variable environments.
Thus, the characteristics of the environment can be assumed to be unknown
and changing significantly dependent on the given task. Besides environmental
uncertainties, various system uncertainties, being the result of imprecise position of workpiece, varying stiffness of the environment, robot tool and robot
itself, etc., have an essential influence on the system's behaviour. For example, in the case of using fixed position/force gains for conventional compliance
control tasks, these controllers perform satisfactorily when the environment
parameters, such as stiffness, are known. However, the same controllers typically exhibit a sluggish response in contact with a softer environment, and
become unstable in contact with a stiffer environment. In this case, one of the
most delicate problems in compliant motion control of the robots interacting
with dynamic environment is the stability of both desired motion and interaction forces. A multitude of various control approaches such as hybrid control,
stiffness control, impedance control, damping control, etc. [362], point to the
stability of the control task as a problem which has not yet been satisfactorily

solved, both from the theoretical and the practical standpoint. In order to extend the problem solving to the more general case when the environment exhibits a dynamic behaviour [59], Vukobratovic and Ekalo have established a unified approach to simultaneously control position and force in an environment with completely dynamic reactions [356]. These authors especially focused their attention on the role of dynamic environment and position stabilization when asymptotic stability of the contact force was ensured. This task is a basic issue in the control of the robot interacting with its dynamic environment. However, without knowing the environment model with sufficient accuracy it is not possible to determine, for instance, the nominal (desired) contact force. Moreover, the insufficiently accurate environment dynamics model can significantly influence the contact task execution. It is evident that in order to overcome these problems, the controller must be capable of adapting its parameters, and possibly its structure, to the changes in the environment parameters and environment model structure. In the case of unknown environment, it is also very difficult to determine the maximum boundary of the position/force feedback gains.

As a result of the mentioned facts, efficient compliance control algorithms must include new features, which are necessary for active compensation of system uncertainties and for the determination of optimal control parameters. It is the *learning feature* that can significantly enhance robotic performance by learning capabilities, which use an a priori low level of information about the model of manipulation robot and environment. Another important characteristic of most contact tasks is their repetitive nature, which is very important for the process of learning by the trial-and-error procedure.

Along with recent extensive research in the area of learning robot control for non-contact tasks, various learning algorithms for compliance tasks have been proposed using different approaches such as iterative-analytical, tabular, connectionist, and hybrid neuro-fuzzy methods. Two essentially different approaches can be distinguished: one, whose aim is the transfer of human manipulation skills to robot controllers and the other, in which the manipulation robot is examined as an independent dynamic system in which learning is achieved through repetition of the working task.

The principle of transferring human manipulation skill has been developed in the papers of Asada and coworkers [15],[14], [17],[16],[389]. The approach is based on the acquisition of manipulation skills and strategies from human experts and subsequent transfer of these skills to robot controllers. It is essentially a playback approach, where the robot tries to execute the working task in the same way as an experienced worker. Various methods and techniques have been evaluated for the acquisition and transfer of human skills to robot controllers. In [15], measurement data of the tool tip position and force exerted on the tool tip in the process of acquisition of human skill are interpreted and translated into a desired strategy through a special program for robot controllers.

It can be said that the generated strategy represents a desired robot trajectory of the tool tip and a desired force profile on it during the working task. This approach was limited exclusively to the operation of robot palletizing. In [14], a method is described for the identification of impedance parameters in the deburring process on the basis of recording data about the worker holding the tool top. These identification parameters were used as input data for the robot controller with the aim to make the robot accomplish the same impedance as the human worker. Of course, this method can be efficient only if the conditions of the working task related to the shape of scraping and stiffness of the working object are similar to those in the preceding process of identification. If the working conditions are time-varying during the task execution, the fixed impedance control cannot give satisfactory results. A further development of the basic idea [17] is the use of pattern recognition techniques for data extraction from the sensory system. These extracted data are the basis for correlation with the control action. The control law of an experienced worker is defined through a set of IF-THEN rules, which have as result a change of the reference trajectory and reference force. Similar ways of demonstrating a compliant motion task by special teaching device for skill acquisition were proposed by Koeppe, Breidenbach and Hirzinger [189]. Learning the mapping of sensor and robot state data to robot motion commands was achieved by rule-based neural networks.The velocities and force signals are input to the network in the form of fuzzy variables. Koeppe showed that velocities should be included to solve the correspondence problem between the human's haptic perception and the motion correction.

The transfer of human skill to the robot controllers described in [16] was accomplished through specialized knowledge that was integrated into the weighting factors of the neural network (Figure 6.1). Teaching of the neural network is based on an off-line procedure using a standard back-propagation algorithm. The proposed neural controller was analyzed for a robotic deburring operation, whereby the input to the three-layer perceptron represents characteristics of the deburring process (characteristics of scraping and tool), while the output of the neural network is the cutting force in normal directions and damping and stiffness system gain. However, no experimental analysis was given, nor direct connections with the real-time control action. In the last paper on this approach [389], linguistic control rules are used to transfer human skills to the grinding process. In real-time control, the working performance was observed and, on the basis of the previously acquired human skills, interpreted through linguistic rules, to accomplish direct adjustment of reference trajectory and tuning of the control gains.

This approach is very interesting and important, although there are some critical issues related to explicit mathematical description of human manipulation skill because of the presence of subconscious knowledge, inconsistent,

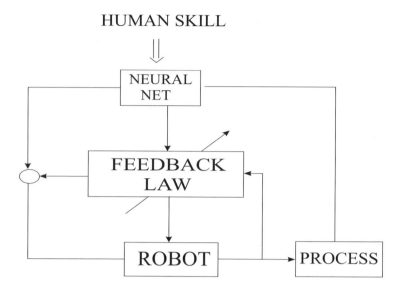

Figure 6.1.   Transfer of human skills to robot controllers by neural network approach

contradictory and insufficient data. These data may cause system instability and wrong behaviour of the robotic system. As is known, the dynamics of the human arm and the robot arm are essentially different, and it is therefore not possible to apply in the same way human skill to the robot controllers. The sensor system for data acquisition of human skill may be insufficient for extracting the complete set of information necessary for transfer to robot controllers. Also, this method is inherently an off-line learning method, whereas, for robot contact tasks, on-line learning is a very important process because of the high-level of robot interaction with the environment and because of unpredictable situations that were not captured in the skill acquisition process. To deal with inconsistent data and human deficiencies, Kaiser and Dillmann [159] have used an approach that takes into account the characteristics of human-generated data, the existence of irrelevant perception and action components and robustness requirements. Initial skill learning was achieved by RBF networks, while skill refinement was based on Gullapalli's stochastic real-valued (SRV) units [106] and reinforcement learning method. The method shows promise, but has a lengthy training process.

Using a similar approach, Skubic and Volz [308], [309], [311], [310] focus on the problems of learning low-level force-based assembly skills from human demonstrations. To avoid position dependencies, force-based discrete states are used to describe qualitatively how the contact with the environment is being made. Sensorimotor skills are modelled using a hybrid control method, which provides a mechanism for combining continuous low-level force control with higher-level discrete event control. The human teacher demonstrates each single-ended contact formation while force data from force sensors are collected, and these data are used to train a state classifier (fuzzy or neural network classifier). The human teacher then demonstrates a skill, and the classifier is used to extract the sequence of single-ended contact formation and transition velocities that comprise the rest of the skill.

A second group of learning methods, based on autonomous on-line learning procedures with the repetition of the working task, is also evaluated through several algorithms [105],[104], [152],[10],[88],[338]. The main distinction between these algorithms is in the aim of the learning, which is in the first case direct adjustment of control signals or parameters, while in the second case the aim of learning is the building of an internal model of the robotic system with compensation for system uncertainties.

The algorithms with on-line modification of control signal are basically related to automated contact-operation (more precisely, the process of assembly). The control goal in the assembly process is to accomplish the whole set of corrective movements using learning rules in order to achieve valid realization of the working task. Asada [12], [13] considered the problem of nonlinear "compliance" in the process of peg-in-hole insertion. The compliance task is defined as a nonlinear mapping of the measured force and moments on the robot end-effector into corrective velocity movements one defined through the linear and angular velocity of the robot end-effector. The nonlinear mapping was represented by a multilayer perceptron with learning rules defined on the basis of the back-propagation method. This is a very interesting approach, but the learning analysis was realized off-line, using recorded input/output patterns without experimental verification.

A very interesting approach belonging to this group is the one by Gullapali and coworkers [105]. The authors use reactive admittance control for compensating the system's uncertainties by on-line control modification and sensor information. Using a similar approach, the realization of active compliance in [12] is based on nonlinear mapping of the admittance from sensors of position and force in commanding velocity movement. The robotic controller learns this mapping through repetitive trial of peg-in-hole insertion. The learning rules are based on the method of associative reinforcement learning [25],[105]. This method can be characterized as random search and reinforcement learning. In contrast to the supervised learning paradigm, the role of the teacher in reinforcement learning is more evaluative than instructional. The teacher provides input to the learning system with an evaluation of the system performance of the robot task according to a certain criterion. Based on both the input to the learning system and the action it generated for that input, the environment computes and returns an evaluation "reinforcement". Over time, the learning system has to learn to respond with the action that has highest expected evaluation. In order to iteratively improve the evaluation obtained for the action associated with each input, the learning system has to determine how the modifying action affects the ensuing evaluation, for example, by estimating the gradient of the evaluation with respect to its actions. The goal of the direct reinforcement learning algorithm is to compensate for the system uncertainties and sensor noise using learning by position/force feedback. The structure of the neural network used is given in Figure 6.2. The neural network has six inputs from the robot sensor system (the peg position X, Y, θ from the sensed joint position and force, and the moments F_x, F_y, M_z from the robot end-effector). In addition to two hidden layers with 15 neurons, the output layer contains 3 stochastic real-valued (SRV) neural units [105] for generation of the linear and angular velocities v_x, v_y and ω_z. These units generate real-valued output stochastically and use the ensuing evaluation to adjust the output so as to maximize the expected evaluation over time [105]. The units do this by estimating the local gradient of the evaluation with respect to its output and using this estimate to perform gradient ascent. Using the SRV units in the output layer enables the network to conduct a search in the space of control actions in order to discover appropriate compliant behaviour. The authors have experimentally verified this approach in robot control with satisfactory results, even in the presence of a high degree of noise and uncertainties.

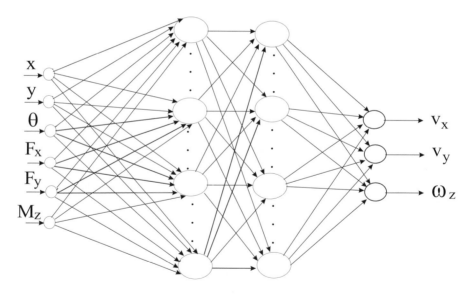

Figure 6.2.   Neural network used for peg-in-hole insertion

Lee and Kim [209] used a special expert system for learning peg-in-hole insertion, starting with the initial rule base. This expert system is implemented as a symbolic system with random searching among the expert's rule in order to discover the best control action in the presence of uncertainties and noise. The result of this approach is similar to the previous one.

In the area of learning impedance control , Cohen and Flash [54] evaluated the appropriateness of reinforcement learning for control in a robot surface wiping task. It is a stochastic scheme of learning, without using models of the robot mechanism and environment. The associative searching network supplies the controller with the input parameters of the target impedance. After a new robot action, the impedance robot controller sends from the sensory system to the associative searching network the values of the end-effector coordinates and a special signal, which represents the measure of compliance between the realized action and desired context. In the learning phase, the network realizes searching that maximizes the value of the special signal for each of the network inputs by changing the weighting factors between the input and output of the network. A unique characteristic of this network is the generation of output vectors using scalar feedback from the robot environment. The reinforcement learning by associative search network might be used for direct adjustment of control signals or parameters but it is not so efficient as the methods for building internal robot models.

For machining operations such as polishing, deburring and grinding, some well-known iterative learning methods from position control are applied for contact tasks [200], [201], [152], [1], [2], [10]. These methods use in the repetitive trials various forms of position and force errors from the previous control process with the aim of bettering robotic performance. As an example we can describe the following control algorithm [10]:

$$u_{k+1} = u_k - \Phi \dot{r}_k + \psi J_\varphi^T(q_k) \Delta f_k \qquad (6.1)$$

where k denotes the number of learning epochs; u_k is the control signal based on some well-known laws for position/force control; $r_k(t) = q_k(t) - q^0(t)$ is the position error from the previous learning epoch k; $f_k = F_k(t) - F^d(t)$ is the force error from the preceding learning epoch k; $J_\varphi(q_k)$ is the Jacobian matrix of the system; Φ and ψ are the gain matrices.

The algorithms of iterative learning use various forms of position and force errors in order to achieve best convergence results, which are in some cases accomplished with restrictive assumptions. Most of the algorithms have been experimentally verified, but one of the main drawbacks of this approach is the poor generalization properties.

From the domain where a primary aim is the learning of internal robot models with compensation for the system uncertainties, we can mention some connectionist algorithms for the learning control of contact tasks [88],[108], [338], [272],[152],[10],[179],[56], [229], [404], [132], [136].

Fukuda and his coworkers [88] proposed a complex control scheme which uses a centralized connectionist structure in the frame of explicit hybrid control laws [380]. Tao and Luh [338] used a technique of model reference control as a basic control algorithm and a neural network as a robust controller in order to compensate for the uncertainties of the dynamic model of multiple robots with redundancy. Kiguchi and Fukuda [179] proposed an adaptive neural controller for hybrid position/force control of a robot in contact with an unknown environment. They used a new type of artificial neurons, which possess visco-elastic properties. The use of the proposed visco-elastic neurons enables the damping of unexpected overshooting and oscillation caused by unknown unmodelled robot or environment dynamics. In their papers [56], [229], [404], the authors have presented a position-based neural force control algorithm for solving a wide range of surface tracking tasks for robot manipulators, involving defined contact with moving rigid objects. This approach comprises a hybrid position/force controller where the neural control algorithm performs force control by modifying the desired joint angles in the force direction. These are fed into a computed torque controller, where the inverse dynamics of the manipulator is represented by neural networks.

A new, robust learning controller for simultaneous position and force control of uncertain constrained manipulators is presented. Using the models of manipulator dynamics and environmental constraint, a task-space reduced-order position dynamics and an algebraic description for the interacting force between the manipulator and its environment are constructed [136]. Based on this treatment, the robust nonlinear control approach and direct adaptive neural network technique are then integrated together. The role of the neural network devices is to adaptively learn those manipulators' structured/unstructured uncertain dynamics as well as the uncertainties with environmental modelling. Then, the effects on tracking performance attributable to the approximation errors of the neural network devices are attenuated to a prescribed level by the embedded nonlinear control. Whenever the adopted NN devices have the potential to effectively approximate those nonlinear mappings to be learned, then this new control scheme can be ultimately less conservative than its counterpart only position/force tracking control scheme. It is believed that this suboptimal position/force tracking control is a novelty in the robotic control literature. In the adaptive neural network position/force tracking controller, the parametric uncertainties and disturbances in manipulator dynamics as well as the geometry and stiffness uncertainties in environmental modelling are capable of being learned by its NN devices. The effects of the approximation error of neural network devices are then attenuated to a prescribed level by the embedded nonlinear controller. Moreover, the neural network devices in this controller do not require any time-consuming off-line training phase. Finally, a simulation study for a constrained two-link planar manipulator is given. Simulation results indicate that the proposed adaptive NN position/force tracking controller performs better in both force and position tracking tasks than its counterpart position/force tracking control scheme.

2. Fundamentals of Connectionist Control Synthesis

The synthesis of new advanced learning control algorithms will be mostly based on the application of connectionist architectures for fast on-line learning of robot dynamic uncertainties used on the executive hierarchical control level in robotic contact tasks. The main idea is connected to the extension and generalization of the approach developed for connectionist control in robotic non-contact tasks in order to deal with the problem of performing position and force control of robot manipulators [359], [165], [168]. The main feature of the proposed hybrid learning control algorithms is the integration of nonrecurrent or recurrent four-layer perceptrons into non-learning control laws (algorithms for stabilization of robot motion and interaction force [355] or impedance control algorithms [121]). The proposed neural network plays the role of a robust controller used to compensate only for the uncertainties of the dynamic model of manipulation robots in contact with dynamic environment, instead of using a "black-box" connectionist approach to modelling the whole dynamic system. The four-layer perceptron, being part of a hybrid learning control algorithm through the process of synchronous training, uses fast learning rules and available sensor information to improve the robot performance progressively, in the minimum possible number of learning epochs. Beside frequently used feedforward perceptrons, the recurrent neural network shows great potential fot handling nonlinear dynamic systems with a high level of system uncertainty [8]. This hybrid approach, founded on the model-based method and connectionist learning, is chosen because of the fact that to some extent information about the dynamic model is always available in the process of control synthesis. In this way, the tracking performance of position and force can be significantly improved using the nonrecurrent or recurrent neural network trained with fast on-line learning algorithms. The basic concept of feedback error learning is adopted as a method for on-line training of the proposed neural structure.

In the text to follow, dynamic modelling of the robot mechanism and robot environment together with the specification of the learning control task, will be presented. After that, the fundamentals of non-learning stabilization and impedance control law as a case of the unified position/force control for contact tasks are described. The learning control algorithms, based on nonrecurrent or recurrent connectionist structures are given along with the basic principles of the proposed connectionist learning methods. Finally, some simulation examples of a robotic deburring process are presented,to verify the effectiveness of the proposed learning algorithms.

2.1   Model of the robot interacting with dynamic environment - task setting

In the case of contact robot tasks, the dynamics of the robot mechanism interacting with its dynamic environment has a crucial effect on the system performance. Hence, it is very important to accurately describe the model of the robotic system dynamics and the model of the environment in the process of control law synthesis.

A model of the robot dynamics interacting with the environment is described by vector differential equations in the form:

$$P = H(q)\ddot{q} + h(q, \dot{q}) + J^T(q)F \tag{6.2}$$

where $P \epsilon R^n$ is the vector of driving torques or forces; $H(q, \theta) : R^n \times \theta \Rightarrow R^{n \times n}$ is the inertia matrix of the system; $h(q, \dot{q}, \theta) : R^n \times R^n \times \theta \Rightarrow R^n$ is the vector which includes centrifugal, Coriolis and gravitational effects; $J(q) : R^n \Rightarrow R^{n \times m}$ is the Jacobian matrix connecting velocities of the robot end-effector and the velocities of robot generalized coordinates; $F \epsilon R^m$ is the vector of generalized forces or of generalized forces and moments from the environment acting on the end-effector; $q \epsilon R^n$ is an n-dimensional vector of the robot generalized coordinates; n is the number of DOFs; m is the number of interacting force components. Here, it will be assumed that $n = m$ (in general $n > m$).

The dynamic model (6.2) can be transformed into an equivalent form in the operational space. This form of the model describes the motion of the end-effector in Cartesian space:

$$\Phi = \Lambda(x)\ddot{x} + \mu(x, \dot{x}) + F \tag{6.3}$$

where the relationships between the corresponding matrices and vectors from equations (6.2) and (6.3) are given as:

$$x = f(q) \tag{6.4}$$

$$\dot{x} = J(q)\dot{q} \tag{6.5}$$

$$\Lambda(x) = J^{-T}(q)H(q)J^{-1}(q) \tag{6.6}$$

$$\mu(x, \dot{x}) = J^{-T}(q)h(q, \dot{q}) - \Lambda(x)\dot{J}(q)\dot{q} \tag{6.7}$$

$$\Phi = J^{-T}(q)\tau \tag{6.8}$$

The end-effector position and orientation are presented by the vector of external coordinates:

$$x = [x_k \ y_k \ z_k \ \varphi \ \theta \ \psi]^T \tag{6.9}$$

where (x_k, y_k, z_k) are the Cartesian coordinates of the reference coordinate frame attached to the manipulator base. Orientation of the end-effector with respect to the base frame is described in terms of the Euler angles (φ, θ, ψ).

The working environment model represents one of the most complex and least investigated issues in robot contact tasks. In the case where the environment does not exhibit displacements that are independent of the robot's motion, a mathematical model of the environment can be described by nonlinear differential equations [59]:

$$M(s)\ddot{s} + L(s, \dot{s}) = F, \qquad s = \varphi(q) \tag{6.10}$$

where s is a vector of the environment coordinates (displacements); $\varphi(q)$ is a vector function connecting two coordinate frames. In the frame of robot joint coordinates, the model of environment dynamics can be presented in the form:

$$M(q)\ddot{q} + L(q, \dot{q}) = S^T(q)F \tag{6.11}$$

where $M(q) \epsilon R^{n \times n}$ is a nonsingular matrix; $L(q, \dot{q}) \epsilon R^n$ is a nonlinear vector function; $S^T(q) \epsilon R^{n \times n}$ is the matrix with $rank(S) = n$. In the case of contact, we shall also assume that all the mentioned matrices and vectors are continuous functions of their arguments.

The presented forms of the robot dynamics model and model of the environment, which can be used for learning control synthesis, have important features that are given in the general nonlinear form of generalized coordinates, although the mathematical models that are commonly used for contact tasks are based on linearized models and external coordinates. Hence, it is convenient to adopt in practice a simplified model of the environment, taking into account the dominant effects, such as stiffness:

$$F = K'(x - x_0) \tag{6.12}$$

or environment damping during the tool motion:

$$F = B' \dot{x} \tag{6.13}$$

where $K' \epsilon R^{n \times n}, B' \epsilon R^{n \times n}$ are the semi-definite matrices describing the environment stiffness and damping, respectively and $x_0 \epsilon R^n$ denotes the coordinate vector of the point of impact between the end-effector (tool) and the constraint surface. However, a more exact approach is to adopt the relationship defined by specification of the target impedance [121]:

$$F = M'\Delta\ddot{x} + B'\Delta\dot{x} + K'\Delta x \tag{6.14}$$

where

$$\Delta x = x - x_0 \tag{6.15}$$

and M' is a positive definite inertia matrix. The matrices M', B', K' define the target impedance which can be selected to correspond to various objectives of the given manipulation task [202]. Obviously, high stiffness is selected in the directions where the environment is compliant and positioning accuracy is important. Low stiffness is selected in the direction where small interaction forces must be maintained. Large values of the matrix B' are specified when energy is to be dissipated, and the matrix M' is used to provide smooth transient behaviour in the system response during the contact. This model represents the basis for the analysis and design of the learning impedance control strategy. The application of this linear model may be limited to the class of contact tasks for which either the environment dynamics can be described sufficiently exactly by a linear equation, or with tasks where the application of a linearized equation of the environment dynamics is admissible. The latter is possible, e.g., in the case of solving assembly tasks, when jamming of the parts during assembly should be avoided. However, the general strategy of impedance control is acceptable without major drawbacks.

The general formulation of a robot control task can be given as the robot motion along the desired trajectory $q_d(t)$ while the desired force $F_d(t)$ is acting between the robot and the environment. In this case, it is important to notice that the desired robot motion $q_d(t)$ and the desired interaction force $F_d(t)$ cannot be arbitrary ones. These two functions must satisfy the following relation:

$$F_d(t) = f(q_d(t), \dot{q}_d(t), \ddot{q}_d(t)) \tag{6.16}$$

The goal of robot learning control in contact tasks can be formulated by the following goal conditions:

$$q^k(t) \rightarrow q_d(t) \qquad\qquad F^k(t) \rightarrow F_d(t) \tag{6.17}$$

where k is the number of learning epochs when the quality of transient response is specified in advance.

2.2   Synthesis of non-learning control algorithms for robotic contact tasks

The principle of control law synthesis on the basis of the preset quality of transient responses [355] is adopted. The control algorithms, as distinct from the control laws synthesized by using familiar traditional approaches, are characterized by the exponential stability of closed-loop systems and ensure the preset quality of transient responses of the motion and interaction force. This principle is important because some of the requirements defined by transient response are obligatory (the real robot motion should converge to the programmed one), while others are desirable. However, simultaneous stabilization of perturbed robot motion and perturbed interaction force with independent requirements for the desired quality of their transient response is not possible, because of the strong connection between the desired force and desired robot motion defined by equation (6.16). Hence, we can divide the synthesis of learning control into two classes: a) the synthesis of stabilization of robot motion with a preset quality of transient processes; b) the synthesis of stabilization of interaction force with a preset quality of transient processes.

Another basic concern in the control synthesis is the impedance control strategy [121]. The fundamental philosophy of *impedance control strategy* is that the manipulator control system should be designed not to track the motion or force trajectory alone, but rather to regulate the mechanical impedance of the manipulator. The impedance control specification consists of a desired dynamic relationship between the Cartesian position errors of the end-effector and the measured end-effector applied force. This approach is a unifying one in the sense that the same control method is used for free motion and for the different types of contact tasks.

Impedance control is mainly used in the contact tasks where precise positioning is needed. This control should ensure: (i) precise positioning; (ii) control of the contact forces; (iii) desired behaviour of the end-effector as a system with the corresponding inertial characteristics, i.e., as a system with a desired mass; (iv) control of the entire dynamic relation between the position of the robot's end-effector and the contact forces; (v) taking into account dynamic characteristics of the robot; (vi) performing the prescribed task regardless of variations of the robot parameters, environment and controller.

The focus of the original approach [121] is on the characterization and control of the dynamic interaction based on manipulator behaviour modification. In this sense, the impedance control is an augmentation of the position control. Several techniques with and without force feedback for modulating the end-point impedances of a general nonlinear manipulator have been proposed. The proposed control design strategy is to adopt the robot behaviour as the inverse of the environment. It means that if the environment behaves like an admittance, impedance control should be applied and vice versa. A control algorithm, referred to as a generalized impedance control, by introducing a general relation between the position and the force errors including force derivatives, has been developed in [210]. Two basic impedance control structures: position-based and torque-based, have been compared in [202]. It has also been stated that the force-based approach is better suited to provide the lower stiffness and damping needed for reducing contact forces. However, with industrial robotic systems, which are primarily designed to realize positioning tasks, the performance of the force-based impedance control is comparatively worse.

2.2.1   Synthesis of non-learning algorithms stabilizing robot motion

Let us synthesize the control law $u(t)$ in such a way as to ensure a desired quality of the robot motion. For this purpose we can specify a family of transient responses given by the vector differential equation [355]:

$$\ddot{\eta} = N(\eta, \dot{\eta}) \qquad\qquad \eta(t) = q(t) - q_d(t) \qquad\qquad (6.18)$$

where N is an n-dimensional vector function. The function N is chosen so that the system (6.18) as a whole is asymptotically stable:

$$N(\eta, \dot{\eta}) = -KP\eta - KD\dot{\eta} \qquad\qquad (6.19)$$

Using the previous equation, the system (6.18) now has the following form:

$$\ddot{\eta} = -KP\eta - KD\dot{\eta} \qquad\qquad (6.20)$$

where $KP\epsilon R^{n\times n}$ is the diagonal matrix of position feedback gains and $KD\epsilon R^{n\times n}$ is the diagonal matrix of velocity feedback gains. The values of these matrices can be chosen according to algebraic stability conditions.

Taking into account the model of the robot in contact with the environment (6.2), one of the possible controlled driving torques has the form [355]:

$$P = H(q)[\ddot{q}_d - KP\eta - KD\dot{\eta}] + h(q, \dot{q}) + J^T(q)F \qquad (6.21)$$

According to the integral model of the robotic system (model of the robot mechanism with the model of robot actuators), the control variable is defined by the following equation:

$$u = f_u(P) \qquad\qquad (6.22)$$

where $u\epsilon R^n$ is the control input and f_u is the nonlinear function that describes the nature of the robot actuator model.

By substituting (6.21) into the robot dynamics model (6.2), we obtain the following equation of the closed-form system:

$$H(q)[\ddot{q} - \ddot{q}_d + KP\eta + KD\dot{\eta}] = 0 \qquad\qquad (6.23)$$

Due to the property of the function N that asymptotic stability is ensured in the whole of the system (6.18), it follows that the control goals

$$\eta(t)_{t\to\infty} \to 0, \quad \dot{\eta}(t)_{t\to\infty} \to 0, \quad \ddot{\eta}(t)_{t\to\infty} \to 0 \qquad (6.24)$$

are achieved.

The proposed control law represents a special version of the well-known computed torque method for contact tasks that does not have learning properties and which uses the available on-line information from the position, velocity and force sensors.

2.2.2 Synthesis of non-learning algorithms stabilizing the interaction force

Let us now consider the second class of stabilizing tasks - stabilization of the interaction force of the robot in contact with the environment. In this case, a strong influence on the stability of the proposed control algorithms is exerted by the type of robot contact tasks described by a dynamic model of the robot environment. Analogously, for this purpose, it has been assumed that the interaction force in transient regime should behave according to the following differential equation:

$$\dot{\mu}(t) = Q(\mu) \tag{6.25}$$

$$Q(\mu) = -KFP\mu - KFI \int_{t_0}^t \mu dt \tag{6.26}$$

where $\mu(t) = F(t) - F_d(t)$, Q is a continuous vector function, such that the asymptotic stability is ensured; $KFP \epsilon R^{n \times n}$ is the matrix of proportional force feedback gains; $KFI \epsilon R^{n \times n}$ is the matrix of integral force feedback gains. When describing the quality of the transient response with respect to perturbation, we use an equivalent relation:

$$\mu(t) = \mu_d(t) + \int_{t_0}^t Q(\mu(\omega))d\omega \tag{6.27}$$

On the basis of (6.25)-(6.26),(6.2), (6.11) we can propose a possible stabilizing control law (computing the value of controlled torque) [355]:

$$P = H(q)M^{-1}(q)[-L(q,\dot{q}) + S^T(q)F] + h(q,\dot{q}) + J^T(q)\{F_d - \int_{t_0}^t [KFP\mu(\omega) + KFI \int_{t_0}^t \mu(\omega)dt]d\omega\} \tag{6.28}$$

The control variable u for the whole robotic system is defined in the same way according to equation (6.22). By substituting (6.28) into the robot dynamic model (6.2), we can obtain the following equation of the closed-form system:

$$J^T(q)\{\mu + \int_{t_0}^t [KFP\mu(\omega) + KFI \int_{t_0}^t \mu(\omega)dt]d\omega\} = 0 \tag{6.29}$$

In this way, due to the property of the function Q, the control law (6.28) ensures the required quality of stabilization of the desired interaction force $F_d(t)$.

2.2.3 Synthesis of non-learning impedance control as a specific case of unified position/force control

Let us start with the general impedance control concept established in [121]. The basic idea of this approach consists OF generating such a robot control law in which the closed control system functions in accordance with the differential equation of the form:

$$F = M'(\ddot{x} - \ddot{x}_0) + B'(\dot{x} - \dot{x}_0) + K'(x - x_0) \tag{6.30}$$

where the constant $m \times n$ matrices M', B', K' represent the matrices of inertia, damping and stiffness of the whole interactive system. These matrices can be chosen by the control system designer dependent on the goals to be achieved by the robot in the technological task.

In the robot joint coordinates, equation (6.30) can be written in the form:

$$F = M(q)(\ddot{q} - \ddot{q}_0) + B(q)(\dot{q} - \dot{q}_0) + K(q)(q - q_0) \tag{6.31}$$

where the joint-coordinates dependent $m \times n$ matrices $M(q)$, $B(q)$, $K(q)$ represent the matrices of inertia, damping and stiffness of the whole interactive system, too. Equation (6.31) is derived from (6.30) by transpose Jacobian matrix multiplication and using the known kinematic relations between the Cartesian and joint space frames.

For the sake of simplification, but with no loss in generality, the impedance control will be considered in the coordinate frame q by putting $m = n$. The goal of impedance control is then to ensure the preset impedance of the closed-loop system defined by equation (6.31). Hereby, it is assumed that the environment dynamics model has also been given in the impedance form:

$$F = \tilde{M}'\ddot{x} + \tilde{B}'\dot{x} + \tilde{K}'x \tag{6.32}$$

or in its alternative form in joint space:

$$F = \tilde{M}(q)\ddot{q} + \tilde{B}(q)\dot{q} + \tilde{K}(q)q \tag{6.33}$$

where $\tilde{M}(q)$, $\tilde{B}(q)$, $\tilde{K}(q)$ and $\tilde{M}'$, $\tilde{B}'$, $\tilde{K}'$ are the corresponding sets of the matrices of inertia, damping and stiffness of the environment.

The desired motion and desired interaction force have to satisfy the equation of environment dynamics, i.e.:

$$F_0(t) = \tilde{M}(q_0)\ddot{q}_0(t) + \tilde{B}(q_0)\dot{q}_0(t) + \tilde{K}(q_0)q_0(t), \quad \forall t \geq t_0 \tag{6.34}$$

Let us assume that the task of impedance control has been solved, i.e., that such control τ has been synthesized that the closed-loop control system functions in accordance with (6.31). Assuming that the differences in the environment dynamics matrices for actual and nominal motion are negligible, it follows from (6.33) and (6.34) that the differences between the actual and programmed robot force interaction with the environment are determined by the equation:

$$F - F_0 = \tilde{M}(\ddot{q} - \ddot{q}_0) + \tilde{B}(\dot{q} - \dot{q}_0) + \tilde{K}(q - q_0) \tag{6.35}$$

By subtracting this equation from (6.31), we obtain the functioning of the closed control system to be governed by the equation:

$$F_0 = (M - \tilde{M})\ddot{\eta} + (B - \tilde{B})\dot{\eta} + (K - \tilde{K})\eta \qquad (6.36)$$

where $\eta(t) = q(t) - q_0(t)$. Evidently, the designer of the control system can choose the matrices M, B and K in such a way that the trivial solution of the unperturbed motion:

$$(M - \tilde{M})\ddot{\eta} + (B - \tilde{B})\dot{\eta} + (K - \tilde{K})\eta = 0 \qquad (6.37)$$

is, say, exponentially stable, whereby all its solutions comply with the inequality:

$$max(\|\eta(t)\|, \|\dot{\eta}(t)\|) \leq C\, e^{-\lambda(t-t_0)}\, \sqrt{\|\eta(t_0)\|^2 + \|\dot{\eta}(t_0)\|^2} \qquad (6.38)$$

where, C is a constant and the index $\lambda > 0$.

In that case, an arbitrary solution of the differential equation (6.36) with the perturbation $F_0(t)$ will be stable, whereby for the estimation of stabilization accuracy the following holds:

$$max(\|\eta(t)\|, \|\dot{\eta}(t)\|) \leq C\, e^{-\lambda(t-t_0)}\, \sqrt{\|\eta(t_0)\|^2 + \|\dot{\eta}(t_0)\|^2} + \frac{C\,C_F}{\lambda} \qquad (6.39)$$

where C_F is a constant that bounds the norm of PFI $F_0(t)$ for all $t > t_0$, i.e., $\|F_0(t)\| \leq C_F$. It is clear that for a large λ, or for a small C_F, which is the same, the impedance control satisfactorily solves the task of robot stabilization in its contact with the environment (6.33).

Let us raise the question of those conditions implying the interaction force $F(t)$ to be close to $F_0(t)$, i.e., whether the impedance control has fully solved the task of robot contact with the dynamic environment. Evidently, if $F_0(t)$ is changing slowly in time, and its norm $\|F_0(t)\|$ is small, the answer is positive. In fact, because η and $\dot{\eta}$ are small, it follows from (6.36) that for a small $\|F_0(t)\|$, $\|\ddot{\eta}\|$ will be small too. And, because the connection between the deviation of the force from its programmed value $F(t) - F_0(t)$ and the deviation of motion PM $\eta(t) = q(t) - q_0(t)$ is determined by (6.35), the small η, $\dot{\eta}$, $\ddot{\eta}$ values cause $\mu(t) = F(t) - F_0(t)$ to be small as well. It is easy to see that all these supplementary requirements, related to a small value of C_F and to the slowly changing $F_0(t)$, are explained by the presence of the perturbation $F_0(t)$ on the left-hand side of equation (6.35), describing the behaviour of the closed control system. It is therefore fully justified to introduce a correction into the impedance equation (6.31), by replacing it with the following equation:

$$F - F_0 = M(\ddot{q} - \ddot{q}_0) + B(\dot{q} - \dot{q}_0) + K(q - q_0) \qquad (6.40)$$

Thus, the closed control system will evidently operate according to equation (6.37) without the perturbation $F_0(t)$. In that case the estimation (6.38) will hold, i.e., $\eta(t)$, $\dot{\eta} \to 0$ at $t \to \infty$ and, due to equation (6.37), the relation $\ddot{\eta}(t) \to 0$ at $t \to \infty$ will also hold. Hence, bearing in mind the relation (6.35) it immediately follows that $F(t) \to F_0(t)$ at $t \to \infty$. In this way the correction carried out by the impedance yields an ideal solution of the contact task. Let us now direct our attention to the fact that in cases when the impedance equation is of the form (6.40), an arbitrary impedance control solves the task of attaining the preset impedance of the closed system. Actually, this is a special case of the control law with the suitably chosen function P, in the case when the environment dynamics is being described by a system of differential equations with the coefficients (6.33). This relates to the fact that the realization of the preset impedance goal, described by (6.40), leads to the equation of the closed system in the form of (6.37), which is a special case of the equation of the preset quality of stabilization $\ddot{\eta} = P(\eta, \dot{\eta})$ [356] in the form:

$$P(\eta, \dot{\eta}) = (M - \tilde{M})^{-1}[(\tilde{B} - B)\dot{\eta} + (\tilde{K} - K)\eta] \qquad (6.41)$$

In that case, the robot control law based on impedance equation (6.40) can be given as

$$\tau = H(q)[\ddot{q}_0 + M^{-1}(-B\dot{\eta} - K\eta + F - F_0)] + h(q, \dot{q}) + J^T(q)F \quad (6.42)$$

This relation provides the "ideal" solution of the task of motion and force stabilization. The control law, synthesized on the basis of the impedance control (6.31), i.e., from the equation without correction, has a form analogous to the relation (6.42), if the term $-F_0$ is omitted:

$$\tau = H(q)[\ddot{q}_0 + M^{-1}(-B\dot{\eta} - K\eta + F)] + h(q, \dot{q}) + J^T(q)F \qquad (6.43)$$

This control law has namely been used for the synthesis of the adaptive impedance control of robots and manipulators [174].

3. Synthesis of Learning Stabilizing Control Laws by Connectionist Structures for Contact Robotic Tasks

The major objective of this section is the application of connectionist structures to fast on-line learning of system uncertainties, used as part of the stabilizing control strategies mentioned above. The asymptotic stability of the proposed non-learning position and force control algorithms requires an exact model of the system. Hence, values of the matrix $H(q)$ and vector $h(q, \dot{q})$ in equation (6.2) should be exactly known. In practice, these values may not be precisely available due to various structured-unstructured uncertainties of the dynamic model and/or external disturbances. In this case, multilayer perceptrons as a connectionist structure can have two different roles in the complete control law. The first is related to the compensation of an a priori unknown model of the matrix $H(q)$ and vector $h(q, \dot{q})$ ("black-box" approach) [360], [268], while the second ("hybrid approach") uses an a priori known model with imprecise values of the matrix $H(q)$ and vector $h(q, \dot{q})$ in the synthesis of learning control law. In the second case, the role of the connectionist structure has a broader sense, because its aim is to compensate for possible uncertainties and differences between the real robot dynamics and the assumed dynamics defined by the user in the process of control synthesis. Another reason to choose this hybrid approach is because the information about the dynamic model is to some extent always available in the process of control synthesis. Thus the process of learning is faster than in the "black-box" case.

In order to achieve good tracking performance in the presence of model uncertainties, a fixed, nonrecurrent, multilayer perceptron is integrated into the motion stabilizing control law, with a desired quality of transient response (Figure 6.3):

$$P = \hat{H}(q)[\ddot{q}_d - KP\eta - KD\dot{\eta}] + \hat{h}(q, \dot{q}) + J^T(q)F + P^{NN} \qquad (6.44)$$

$$P^{NN} = F_1(w_{jk}, q_d, \dot{q}_d, \ddot{q}_d) \qquad (6.45)$$

where F_1 is a nonlinear mapping for the multilayer perceptron NN; P^{NN} is a compensation part of the learning control law; w_{jk} are the weighting factors for the perceptrons NN; P is a controlled driving torque; $\hat{H}(q)$ and $\hat{h}(q, \dot{q})$ are assumed functions of the robot dynamics model.

In a similar way, the fixed nonrecurrent multilayer perceptron is integrated into the non-learning control law with the desired quality of transient process for the interaction force (Figure 6.4):

$$P^{NN} = F_2(w_{jk}^{NNab}, q_d, \dot{q}_d, \ddot{q}_d) \qquad (6.46)$$

$$P = \hat{H}(q)\hat{M}^{-1}(q)[-\hat{L}(q, \dot{q}) + \hat{S}^T(q)F] + h(q, \dot{q}) +$$
$$J^T(q)\{F_d - \int_{t_0}^{t}[KFP\mu(\omega) + KFI\int_{t_0}^{t}\mu(\omega)dt]d\omega\} + P^{NN}$$

$$(6.47)$$

where F_2 is a nonlinear mapping for the perceptron NN; P^{NN} is the compensation part of the learning control law; w_{jk} are weighting factors for the perceptron NN; $\hat{M}(q), \hat{L}(q, \dot{q}), \hat{S}(q)$ are assumed functions of the robot environment model.

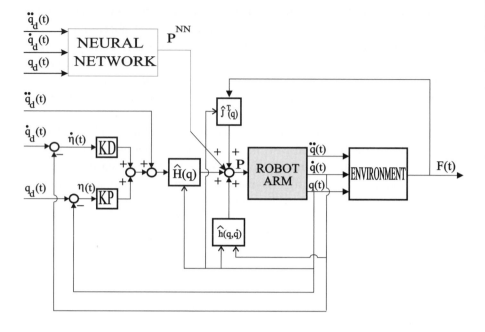

Figure 6.3.   Scheme of learning connectionist law stabilizing robot motion

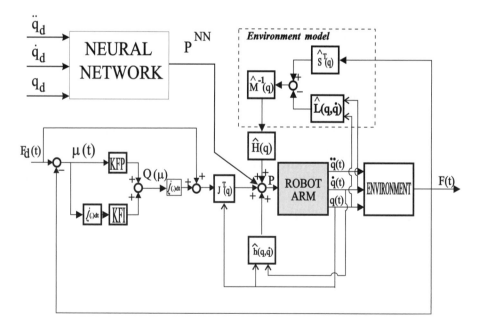

Figure 6.4.   Scheme of learning connectionist law stabilizing interaction force

The integration of connectionist structures ensures the necessary dynamic compensation within the learning position and force control laws. In the first case, compensation by the neural network is connected with uncertainties of the robot dynamic model, while in the second case, compensation is connected with uncertainties of the robot dynamic model and environment dynamics. In the proposed approach, the neural networks compensate for the general uncertainties incorporated in the robot-environment model, while the use of connectionist structures for separate compensation of uncertainties of the dynamics models of the robot and environment represents an interesting issue for further research. The final result of the proposed approach is a trainable robot controller architecture which uses the neural network model as part of the robust stabilizing learning control laws.

3.1 Connectionist structure and on-line learning algorithm

The topology of the proposed connectionist structure for learning robot control is defined by a four-layer perceptron with two hidden layers (Figure 6.5). This topology is chosen in order to avoid the process of normalization of network output signals due to limited values of the output of sigmoid functions. Two hidden layers were used because it was enough to realize the usual mapping functions. The sigmoid function is set as an activation function in both hidden layers, while the activation function for input and output layers is the identity function. There is a bias member in the input and hidden layers.

The neural network NN has an input layer with $3n$ neurons (inputs are $[q_d(t)\dot{q}_d(t)\ddot{q}_d(t)]^T$) and the output layer with n neurons (P^{NN}). One of the main design parameters of the neural network is the number of neurons in each layer. To investigate the effect of neuron size on performance of the learning algorithm, training was conducted for various sizes of the neurons. Hence, based on simulation experiments, the following network topology was adopted: 19 neurons (input layer) - 25 neurons (first hidden layer) - 13 neurons (second hidden layer) - 6 neurons (output layer).

Based on the mentioned facts, general relations in the learning process for forward passage through the four-layer perceptron are given in the form:

$$s_2(t) = W_{12}^T(t-1)i_1(t) \qquad i_{10}(t) = 1 \qquad (6.48)$$
$$o_{2a}(t_k) = 1/(1 + exp(-s_{2a}(t))) \quad a = 1,...,L_1 \quad o_{20}(t) = 1 \,(6.49)$$
$$s_3(t) = W_{23}^T(t-1)o_2(t) \qquad (6.50)$$
$$o_{3b}(t) = 1/(1 + exp(-s_{3b}(t))) \quad b = 1,...,L_2 \quad o_{30}(t) = 1 \;(6.51)$$
$$s_4(t) = W_{34}^T(t-1)o_3(t) \qquad (6.52)$$
$$P_c^{NN}(t) = s_{4c}(t) \quad c = 1,..,n \qquad (6.53)$$

where $s_2(t), s_3(t), s_4(t)$ are the output arrays of the linear part of the network; $o_2(t)$, $o_3(t)$ are the output arrays of the hidden layers; $W_{12} = [w_{12ij3n+1\times L1}]$,

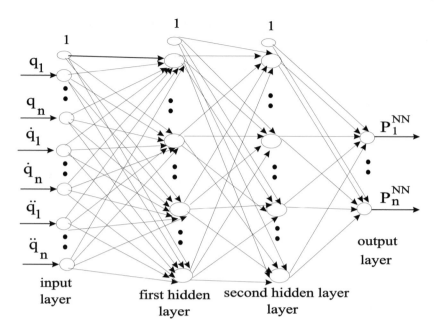

Figure 6.5.   Neural network architecture for compensation of system uncertainties

$W_{23} = [w_{23ij_{L1+1 \times L2}}]$, $W_{34} = [w_{34ij_{L2+1 \times n}}]$ are the matrices of weighting factors between the appropriate layers; $i_1(t)$ is the array of inputs into the network $=[q_d(t)\dot{q}_d(t)\ddot{q}_d(t)]^T$).

Training of the proposed connectionist structure has to be accomplished exclusively in an on-line regime, by using learning rules based on the recursive least square (RLS) algorithm with inverse mapping [168]. The aim of the proposed training algorithm is to define optimal values for the matrices W_{12}, W_{23}, W_{34} using the models of linear systems (equations (6.48),(6.50),(6.52)) and RLS estimation method. In the application of this method a problem arises concerning the specification of desired output values of the linear subsystems. This problem is solved by the inverse mapping of desired output values of the nonlinear subsystems (equations (6.49) and (6.51)), together with the linear mapping algorithm [168] in order to ensure values between 0 and 1 (because of the sigmoid function used as activation function). These learning rules are chosen in order to enhance the learning speed in comparison with standard back propagation algorithm. The applied learning rules, based on the application of the RLS algorithm with inverse mapping, are given by the following relations:

$$\delta^4(t) = \tau^{NN}(t) - P^{NN}(t) \tag{6.54}$$

$$P_{34}(t) = P_{34}(t-1) - \frac{P_{34}(t-1)o_3(t)o_3(t)^T P_{34}(t-1)}{1 + o_3(t)^T P_{34}(t)o_3(t)} \tag{6.55}$$

$$W_{34}(t) = W_{34}(t-1) + P_{34}(t)o_3(t)\delta^{4^T}(t) \tag{6.56}$$

$$o_3^z(t) = o_3(t) + W_{34}(t)\delta^4(t) \tag{6.57}$$

$$o_{3b}^{zpr}(t) = 0.01 + (o_{3b}^z(t) - o_{3min}(t))\frac{1 - 0.02}{o_{3max}(t) - o_{3min}(t)} \quad b = 1, .., L_2 \tag{6.58}$$

$$s_{3b}^z(t) = -\ln\frac{1 - o_{3b}^{zpr}(t)}{o_{3b}^{zpr}(t)} \quad b = 1, .., L_2 \tag{6.59}$$

$$P_{23}(t) = P_{23}(t-1) - \frac{P_{23}(t-1)o_2(t)o_2(t)^T P_{23}(t-1)}{1 + o_2(t)^T P_{23}(t-1)o_2(t)} \tag{6.60}$$

$$W_{23}(t) = W_{23}(t-1) + P_{23}(t)o_2(t)[s_3^z(t) - W_{23}(t-1)^T o_2(t)]^T \tag{6.61}$$

$$o_2^z(t) = o_2(t) + W_{23}(t)[s_3^z(t) - W_{23}(t-1)^T o_2(t)] \tag{6.62}$$

$$o_{2a}^{zpr}(t) = 0.01 + (o_{2a}^z(t) - o_{2min}(t))\frac{1 - 0.02}{o_{2max}(t) - o_{2min}(t)} \quad a = 1, .., L_1 \tag{6.63}$$

$$s_{2a}^z(t) = -\ln\frac{1 - o_{2a}^{zpr}(t)}{o_{2a}^{zpr}(t)} \quad a = 1, .., L_1 \tag{6.64}$$

$$P_{12}(t) = P_{12}(t) - \frac{P_{12}(t-1)i_1(t)i_1(t)^T P_{12}(t-1)}{1 + i_1(t)^T P_{12}(t-1)i_1(t)} \tag{6.65}$$

$$W_{12}(t) = W_{12}(t-1) + P_{12}(t)i_1(t)[s_2^z(t) - W_{12}(t-1)^T i_1(t)]^T \tag{6.66}$$

where P_{12}, P_{23}, P_{34} are the covariance matrices for the RLS algorithm; τ^{NN} is a training torque for the neural network; $o_{2max}, o_{3max}, o_{2min}, o_{3min}$ are maximal and minimal values of the outputs of the nonlinear subsystems $o_2(t)$ and $o_3(t)$.

One of the key issues in designing a neural controller is to obtain the desired training signal, i.e., the desired training torque. Training torques for learning rules are derived based on the user-defined model and on the values from the robot sensors, in order to achieve good tracking performance for simultaneous tracking of position and force. In fact, the training torque τ^{NN} represents the difference between the proposed control laws (6.44) or (6.47), based on the values of the dynamic robot model ar the time instant $t - 1$ and the robust control law for the desired robot model obtained after measuring new values of the system states. This robust control law is given by the following relation:

$$P_{ro}(t) = \hat{H}(t)\ddot{q}(t) + \hat{h}(t) + J^T(t)F(t) \tag{6.67}$$

In the first case of stabilization task, training torque is defined by the following equation:

$$\tau_{NN}(t) = P(t) - P_{ro}(t) = \hat{H}(t-1)[\ddot{q}_d(t) - KP\eta(t) - KD\dot{\eta}(t)]$$

$$+\hat{h}(t-1) - \hat{H}(t)\ddot{q}(t) - \hat{h}(t) \tag{6.68}$$

while in the second case of stabilization task, training torque is defined by:

$$
\begin{aligned}
\tau_{NN}(t) &= P(t) - P_{ro}(t) = \hat{H}(t-1)\hat{M}^{-1}(t-1)[-\hat{L}(t-1) \\
&\quad + \hat{S}^T(t)F(t-1)] + \hat{h}(t-1) \\
&\quad + J^T(t-1)\{F_d(t) - \int_{t_0}^{t}[KFP\mu(\omega) + KFI\int_{t_0}^{t}\mu(\omega)dt]d\omega\} \\
&\quad - \hat{H}(t)\ddot{q}(t) - \hat{h}(t) - J^T(t)F(t) \tag{6.69}
\end{aligned}
$$

4. Synthesis of Connectionist Learning Impedance Laws for Robotic Contact Tasks

The objective of this section is the application of recurrent connectionist structures for fast and robust on-line learning of system uncertainties used as part of the impedance control strategies mentioned above. The role of recurrent neural networks learning connectionist impedance control laws is based on the same principle as in the case of stabilizing learning control laws (compensation of system uncertainties). Beside the frequently used feedforward perceptrons, recurrent neural network shows great potential in handling nonlinear dynamic systems with high levels of system uncertainties. More recently, several studies have noted that an appropriate dynamic mapping may be efficiently realized by a dynamic recurrent network [260],[315]. Dynamic recurrent neural networks may have better prediction capabilities than the static feedforward neural maps. There are many different architectures of recurrent neural networks. In this case, a recurrent neural network architecture is selected that has inner feedback connections between the neurons in hidden layers.

In order to achieve good tracking performance in the presence of model uncertainties, the recurrent neural network is integrated into the impedance control law (Figure 6.6):

$$\tau_u = \hat{H}(q)[\ddot{q}_0 + M^{-1}(-B\dot{\eta} - K\eta + F - F_0)] + \hat{h}(q,\dot{q}) + J^T(q)F + P^{RNN} \tag{6.70}$$

or in the case of a learning version of the algorithm (6.43):

$$\tau_u = \hat{H}(q)[\ddot{q}_0 + M^{-1}(-B\dot{\eta} - K\eta + F)] + \hat{h}(q,\dot{q}) + J^T(q)F + P^{RNN} \tag{6.71}$$

where

$$P^{RNN} = F_1(w_{jk}^{NNab}, q_0, \dot{q}_0, \ddot{q}_0) \tag{6.72}$$

where F_1 is the nonlinear mapping for the recurrent network RNN; P^{RNN} is the compensation part of the learning control law; w_{jk}^{RNNab} are the weighting factors for the recurrent network RNN; w_{kJ}^{RNA} are the weighting factors for the recurrent network RNN between neurons in one hidden layer of the network including auto-feedback connections; τ_u is the controlled driving torque; $\hat{H}(q)$ and $\hat{h}(q,\dot{q})$ are imprecisely defined functions of the robot dynamic model.

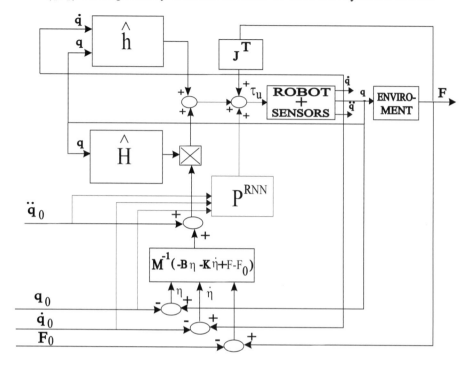

Figure 6.6.   Scheme of learning impedance connectionist law

In this way the integration of recurrent connectionist structures ensures the necessary dynamic compensation of system uncertainties in the impedance control laws. The final result of the proposed approach is a trainable robot controller architecturem which uses the neural network model as part of the robust impedance learning control laws.

4.1   Recurrent connectionist structure and on-line learning rules

The topology of the proposed connectionist structure for learning robot control is defined by a recurrent network with the inner feedback connections and sigmoid function as activation functions in both of the hidden layers. The network RNN has an input layer with $3n$ neurons and an output layer with n neurons. The activation function for input and output layers is the identity function. This form of network with two hidden layers and an output linear layer is chosen because of good representation capabilities and unlimited values of the network outputs. In order to enable best performance of the learning algorithm, the number of neurons in hidden layers is determined by simulation experiments and on the basis of experience. In this case the following network topology is adopted: 19 - 25 - 13 - 6. In the case of the proposed recurrent neural network, general relations in the learning process for forward passage are given in the form:

$$s_2(t_k) = W_{12}^T(t_{k-1})i_1(t_k) + W_{22}(t_{k-1})o_2(t_{k-1}) \quad i_{10}(t_k) = 1 \quad (6.73)$$

$$o_{2a}(t_k) = \frac{1}{1 + exp(-s_{2a}(t_k))} \quad a = 1,...,L_1 \quad o_{20}(t_k) = 1 \quad (6.74)$$

$$s_3(t_k) = W_{23}^T(t_{k-1})o_2(t_k) + W_{33}(t_{k-1})o_3(t_{k-1}) \quad (6.75)$$

$$o_{3b}(t_k) = \frac{1}{1 + exp(-s_{3b}(t_k))} \quad b = 1,...,L_2 \quad o_{30}(t_k) = 1 \quad (6.76)$$

$$s_4(t_k) = W_{34}^T(t_{k-1})o_3(t_k) \quad (6.77)$$

$$P^{RNN}(t_k) = s_4(t_k) \quad (6.78)$$

where $s_2(t_k),s_3(t_k),s_4(t_k)$ are the output arrays of the linear part of the network; $o_2(t_k)$, $o_3(t_k)$ are the output arrays of the hidden layers; $W_{12} = [w_{12ij_{N_u \times L1}}]$, $W_{23} = [w_{23ij_{L1+1 \times L2}}]$, $W_{34} = [w_{34ij_{L2+1 \times 1}}]$ are the matrices of weighting factors between the appropriate layers; $W_{22} = [w_{22ij_{L1 \times L1}}]$, $W_{33} = [w_{33ij_{L2 \times L2}}]$ are the matrices of weighting factors between the neurons in hidden layers including autofeedback connection; $i_1(t)$ is the array of inputs into the network (in this case they are nominal internal coordinates, desired internal velocities, and desired internal accelerations).

Training of the proposed connectionist structure is accomplished exclusively in an on-line regime, by using feedback error learning rules based on the recurrent back propagation algorithm:

$$\delta^4(t_k) = e^{RNN}(t_k) \quad (6.79)$$

$$\nabla^4(t_k) = o_3(t_k)\delta^{4^T}(t_k) \quad (6.80)$$

$$\nabla^{44}(t_k) = o_3(t_{k-1})\delta^{4^T}(t_k) \quad (6.81)$$

$$\delta_b^3(t_k) = o_{3b}(t_k)(1 - o_b^3(t_k)) \sum_{j=1}^{N_y} w_{34bj}(t_{k-1})\delta_j^4(t_k) \quad j = 1,..,L_2+1 \quad (6.82)$$

$$\nabla^3(t_k) = o_2(t_k)\delta^{3^T}(t_k) \quad (6.83)$$

$$\nabla^{33}(t_k) = o_2(t_{k-1})\delta^{3^T}(t_k) \quad (6.84)$$

$$\delta_a^2(t_k) = o_a^2(t_k)(1 - o_a^2(t_k)) \sum_{j=1}^{L_2} w_{23aj}(t_{k-1})\delta_j^3(t_k) \quad j = 1,..,L_1+1$$

$$(6.85)$$

$$\nabla^2(t_k) = u^1(t_k)\delta^{2^T}(t_k) \qquad (6.86)$$

$$W_{34}(t_k) = W_{34}(t_{k-1}) + \eta^{34}\nabla^4(t_k) + \alpha^{34}[W_{34}(t_{k-1}) - W_{34}(t_{k-2})] \qquad (6.87)$$

$$W_{33}(t_k) = W_{33}(t_{k-1}) + \eta^{33}\nabla^{44}(t_k) + \alpha^{33}[W_{33}(t_{k-1}) - W_{33}(t_{k-2})] \qquad (6.88)$$

$$W_{23}(t_k) = W_{23}(t_{k-1}) + \eta^{23}\nabla^3(t_k) + \alpha^{23}[W_{23}(t_{k-1}) - W_{23}(t_{k-2})] \qquad (6.89)$$

$$W_{22}(t_k) = W_{22}(t_{k-1}) + \eta^{22}\nabla^{32}(t_k) + \alpha^{22}[W_{22}(t_{k-1}) - W_{22}(t_{k-2})] \qquad (6.90)$$

$$W_{12}(t_k) = W_{12}(t_{k-1}) + \eta^{12}\nabla^2(t_k) + \alpha^{12}[W_{12}(t_{k-1}) - W_{12}(t_{k-2})] \qquad (6.91)$$

where $\eta^{34}, \eta^{33}, \eta^{23}, \eta^{22}, \eta^{12}$ are the appropriate learning rates and $\alpha^{34}, \alpha^{33}, \alpha^{23}, \alpha^{22}, \alpha^{12}$ are the appropriate momentum factors.

Training torques for learning rules are derived from the user-defined model and values from robot sensors in order to achieve good tracking performance for simultaneous position and force tracking. In the first case of stabilization task, the training torque is defined by the following equation:

$$e^{RNN}(t_k) = \hat{H}(t_{k-1})[\ddot{q}_0(t_k) + M^{-1}(-B\dot{\eta}(t_k) - K\eta(t_k) + F(t_k)$$
$$-F_0(t_k)] + \hat{h}(t_{k-1}) - \hat{H}(t_k)\ddot{q}(t_k) - \hat{h}(t_k) \qquad (6.92)$$

or

$$e^{RNN}(t_k) = \hat{H}(t_{k-1})[\ddot{q}_0(t_k) + M^{-1}(-B\dot{\eta}(t_k) - K\eta(t_k) + F(t_k)]$$
$$+\hat{h}(t_{k-1}) - \hat{H}(t_k)\ddot{q}(t_k) - \hat{h}(t_k) \qquad (6.93)$$

5. Case Study

To verify the proposed learning connectionist algorithms we will describe several simulation examples of the deburring process with industrial robot Manutec r3 (Figure 6.7). The parameters of mechanical structure and actuators were taken from [345]. The technological working demands for this operation are defined by the following statements: a) The working tool of the robot is in the form of a rotational-milling tool, performing surface processing in the plane parallel with a $X - Y$ plane; b) Tool trajectory is 100 $[mm]$ long. c) Task of the robot is to carry out the machining process of the worksurface along the prescribed trajectory with the desired contact force $F_N^0 = 5$ $[N]$ and the prescribed velocity of 25 $[mm/s]$.

A general model of impedance was adopted for the environment model, so that the dynamic environment (i.e., the worksurface) resists the tool motion in all coordinate directions x. The environment model is then represented by the following equation:

$$M\ddot{x}(t) + B\dot{x} + Kx(t) = F(t) \tag{6.94}$$

and in the case of impedance control laws the model is represented by: $F = \tilde{M}'\ddot{x} + \tilde{B}'\dot{x} + \tilde{K}'x$. The corresponding matrices in the environment model were adopted in the form:

Inertia matrix $M = \tilde{M}'$:

$$\tilde{M}' = diag\{\tilde{M}'^i\}, \quad i = 1 \ldots n, \quad \tilde{M}'^1 = 28.173; \quad \tilde{M}'^2 = 101.4240;$$

$$\tilde{M}'^3 = 28.173; \quad \tilde{M}'^4 = 0.1; \quad \tilde{M}'^5 = 0.1; \quad \tilde{M}'^6 = 0.1; \tag{6.95}$$

Damping matrix $B = \tilde{B}'$:

$$\tilde{B}' = diag\{\tilde{B}'^i\}, \quad i = 1 \ldots n, \quad \tilde{B}'^1 = 1061.571; \quad \tilde{B}'^2 = 4458.599;$$

$$\tilde{B}'^3 = 1061.571; \quad \tilde{B}'^4 = 0.1; \quad \tilde{B}'^5 = 0.1; \quad \tilde{B}'^6 = 0.1; \tag{6.96}$$

Stiffness matrix $K = \tilde{K}'$:

$$\tilde{K}' = diag\{\tilde{K}'^i\}, \quad i = 1 \ldots n, \quad \tilde{K}'^1 = 10^4; \quad \tilde{K}'^2 = 10^5;$$

$$\tilde{K}'^3 = 10^4; \quad \tilde{K}'^4 = 0.1; \quad \tilde{K}'^5 = 0.1; \quad \tilde{K}'^6 = 0.1; \tag{6.97}$$

The matrices of the overall interactive closed control system (6.30) for the control law (6.70) are set in the following forms:

$$M' = \begin{bmatrix} 80 & 0 & 0 & 0 & 0 & 0 \\ 0 & 80 & 0 & 0 & 0 & 0 \\ 0 & 0 & 80 & 0 & 0 & 0 \\ 0 & 0 & 0 & 80 & 0 & 0 \\ 0 & 0 & 0 & 0 & 80 & 0 \\ 0 & 0 & 0 & 0 & 0 & 80 \end{bmatrix}; \tag{6.98}$$

$$B' = \begin{bmatrix} 4000 & 0 & 0 & 0 & 0 & 0 \\ 0 & 4000 & 0 & 0 & 0 & 0 \\ 0 & 0 & 4000 & 0 & 0 & 0 \\ 0 & 0 & 0 & 4000 & 0 & 0 \\ 0 & 0 & 0 & 0 & 4000 & 0 \\ 0 & 0 & 0 & 0 & 0 & 4000 \end{bmatrix}; \tag{6.99}$$

$$K' = \begin{bmatrix} 50000 & 0 & 0 & 0 & 0 & 0 \\ 0 & 50000 & 0 & 0 & 0 & 0 \\ 0 & 0 & 50000 & 0 & 0 & 0 \\ 0 & 0 & 0 & 50000 & 0 & 0 \\ 0 & 0 & 0 & 0 & 50000 & 0 \\ 0 & 0 & 0 & 0 & 0 & 50000 \end{bmatrix}; \tag{6.100}$$

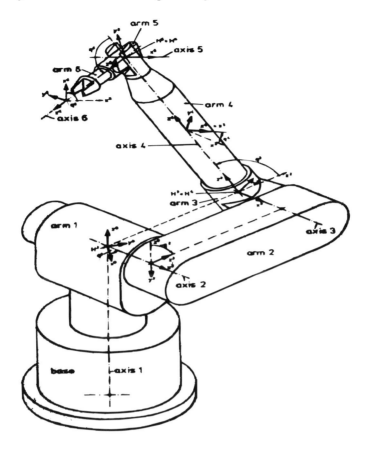

Figure 6.7. Industrial robot MANUTEC r3

while the same matrices for the second control law (6.71) are:

$$
M' = \begin{bmatrix}
127400 & 0 & 0 & 0 & 0 & 0 \\
0 & 127400 & 0 & 0 & 0 & 0 \\
0 & 0 & 127400 & 0 & 0 & 0 \\
0 & 0 & 0 & 127400 & 0 & 0 \\
0 & 0 & 0 & 0 & 127400 & 0 \\
0 & 0 & 0 & 0 & 0 & 127400
\end{bmatrix} ;
$$

$$(6.101)$$

$$B' = \begin{bmatrix} 800000 & 0 & 0 & 0 & 0 & 0 \\ 0 & 800000 & 0 & 0 & 0 & 0 \\ 0 & 0 & 800000 & 0 & 0 & 0 \\ 0 & 0 & 0 & 800000 & 0 & 0 \\ 0 & 0 & 0 & 0 & 800000 & 0 \\ 0 & 0 & 0 & 0 & 0 & 800000 \end{bmatrix};$$

(6.102)

$$K' = \begin{bmatrix} 400000 & 0 & 0 & 0 & 0 & 0 \\ 0 & 400000 & 0 & 0 & 0 & 0 \\ 0 & 0 & 400000 & 0 & 0 & 0 \\ 0 & 0 & 0 & 400000 & 0 & 0 \\ 0 & 0 & 0 & 0 & 400000 & 0 \\ 0 & 0 & 0 & 0 & 0 & 400000 \end{bmatrix};$$

(6.103)

The following initial conditions were used in the simulation: initial deviation of the end-effector tip $\Delta y = +30\,[\mu m]$ and $\Delta z = +50\,[\mu m]$. The robot gripper started with a zero initial velocity. The settling time of the desired contact force is given as $t = 0.5\,[sec]$.

The feedback gains for stabilizing control laws with the required quality of position transient response have been chosen in the form of diagonal matrices:

$$KP = diag\{kp^i\}, \quad kp^i = 8. \quad KD = diag\{kd^i\}, \quad kd^i = 16. \quad i = 1, .., n.$$

(6.104)

The feedback gains for stabilizing control laws with the required quality of force transient response have been chosen in the form of the diagonal matrices:

$$KFP = diag\{kfp^i\}, \quad kfp^i = 8. \quad KFI = diag\{kfi^i\}, \quad kfi^i = 16. \quad i = 1, .., n.$$

(6.105)

In the simulation experiments, the uncertainties of the robot dynamics model and of the dynamic environment were defined by parametric disturbances with approximately 20 % variation from the nominal values of the link mass, moment of inertia and impedance matrices. In the case of the application of the back propagation algorithm for recurrent neural network, learning rates were $\eta^{ij} = 0.05, i = 1, ..3; j = 2, ..4$ and the momentum factors $\alpha^{ij} = 0.9, i = 1, ..3; j = 2, ..4$.

Simulation results for the first class of stabilization tasks (position stabilization) are presented in Figures 6.8 - 6.9. Comparison of the external position error Δz for a non-learning control algorithm with the assumed (inexact) dynamic model of robot mechanism with learning connectionist control algorithm in the first and the fifth learning epoch is presented in Figure 6.8. For the same non-learning and learning control algorithms the error of normal force is shown in Figure 6.9. It is evident that this non-learning control results in large position and force errors due to the use of an inexact model. This was the reason for using learning control laws by neural networks. The results obtained with the neural network approach show that repetitive trials yield a considerable decrease position-tracking errors and force errors, i.e., the robot dynamic learning in the case of a stabilization task with the required quality of transient responses for position, was accomplished.

Exactly the same analysis was performed for the second class of stabilization tasks, the results of which are shown in Figures 6.10-6.11. The conclusions are the same as in the previous case, with somewhat less good qualitative results in the convergence of the learning process, but with an important tendency of the learning process to yield better system performance.

The simulation results for first class of impedance control algorithms (Class I, defined by equations (6.42) and (6.70)) are presented in Figures 6.12 - 6.15. Comparison of the external position errors $\Delta x, \Delta y, \Delta z$ for the non-learning impedance control algorithm with exact and inexact dynamic model of robot mechanism and learning impedance control algorithm is presented in Figures 6.12 - 6.14. Error of normal force in the case of a non-learning and learning control law, with exact and inexact dynamic model of robot mechanism is also shown in Figure 6.15. It is evident that nonlearning control with the inexact model results in large position and force errors. It is the reason for using efficient learning control laws by neural networks. The results presented in these figures show that the position-tracking errors and force errors have been considerably lowered by repetitive trials, i.e., the robot dynamics learning in the case of impedance control laws with learning has been accomplished.

The same analyses were performed for the second class of impedance control laws (Class II, defined by (6.43) and (6.71)), the results of which are shown in Figures 6.16 - 6.19. The conclusion about using connectionist structures in the learning control is the same as in the previous case, but this class of non-learning and learning impedance control laws give worse dynamic features than Class I.

In this chapter we have demonstrated that the problem of tracking a specified reference trajectory with a specified force profile can be efficiently solved by applying the so-called connectionist architectures. The proposed connectionist structures, as integral parts of control strategies for robot contact tasks, enables a fast and robust on-line learning of the system uncertainties. The main feature of the proposed learning robust controller is the utilization of a four-layer nonrecurrent or recurrent neural network to control the position and interactive force in the case of manufacturing contact tasks. In this way connectionist structures counteract the effects of the robot-environment model. The neural network uses a nonrecurrent or recurrent version of back-propagation learning rules and available sensor information in order to achieve best robotic performance for minimal possible number of learning epochs through the process of synchronous training.

Moreover, the reported results show that the proposed learning control schemes are promising in the following aspects. First, by using these algorithms, the on-line learning stabilization of robot motion and interaction force is achieved. Desired tracking performance can be achieved very quickly, after several learning epochs. The proposed connectionist structures are also simple for real-time control. This control technique does not require knowledge of the bounds of uncertainties of the dynamic model. Hence, it can be implemented to control a more general class of dynamic system in the case where the system model is not known.

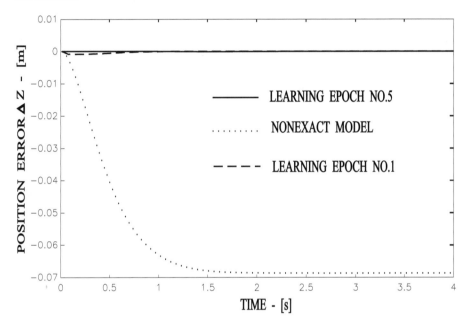

Figure 6.8.   External error ΔZ - position stabilization

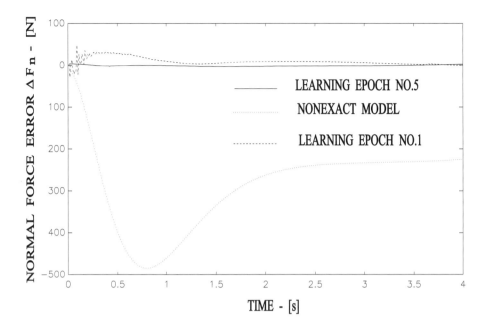

Figure 6.9. Error of normal force - position stabilization

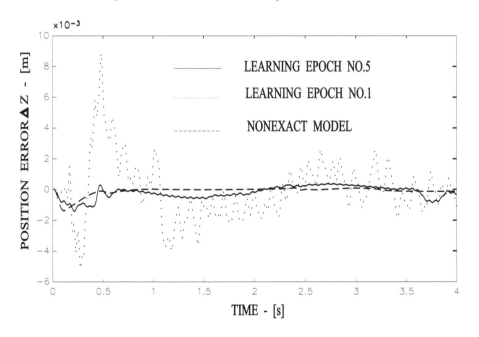

Figure 6.10. External error ΔZ - force stabilization

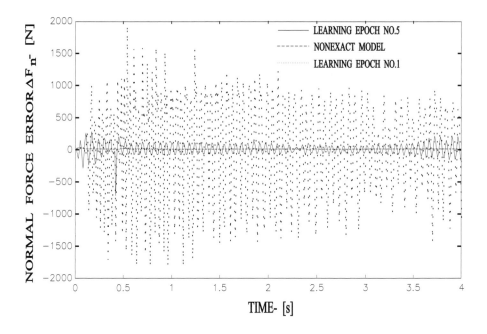

Figure 6.11.   Error of normal force - force stabilization

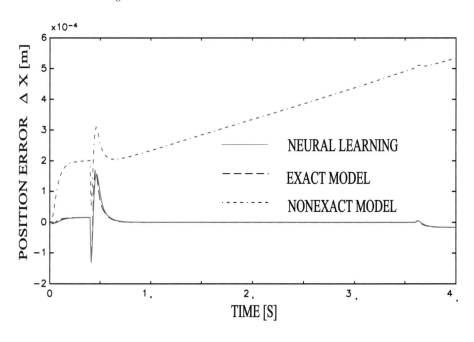

Figure 6.12.   Non-learning and learning impedance control class I - external error ΔX

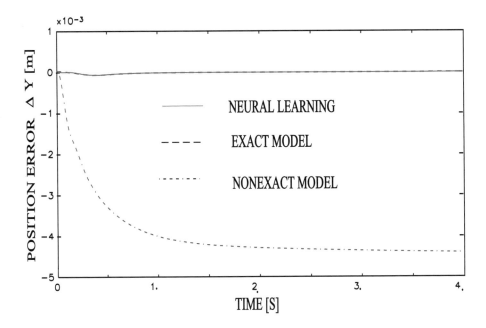

Figure 6.13.   Non-learning and learning impedance control class I - external error ΔY

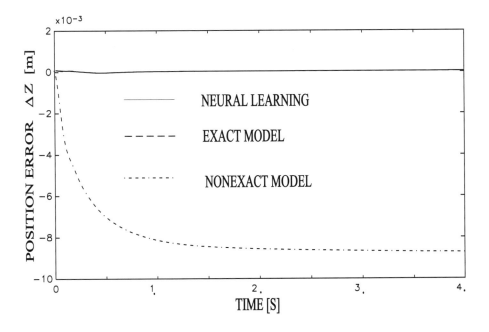

Figure 6.14.   Non-learning and learning impedance control class I - external error ΔZ

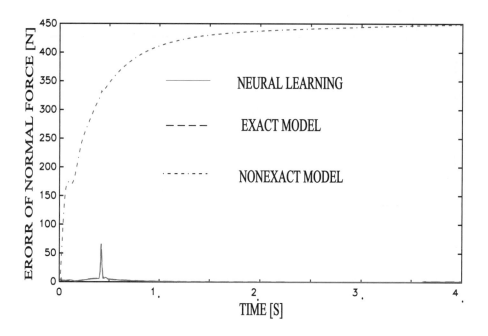

Figure 6.15.   Non-learning and learning impedance control class I - error of normal force

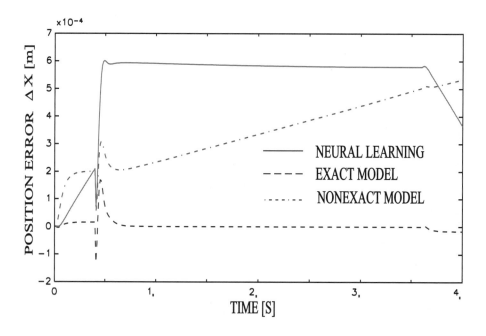

Figure 6.16.   Non-learning and learning impedance control class II - external error ΔX

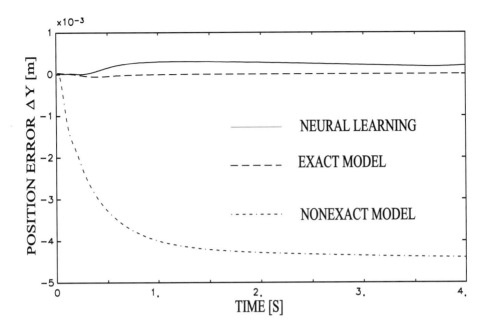

Figure 6.17.   Non-learning and learning impedance control class II -external error ΔY

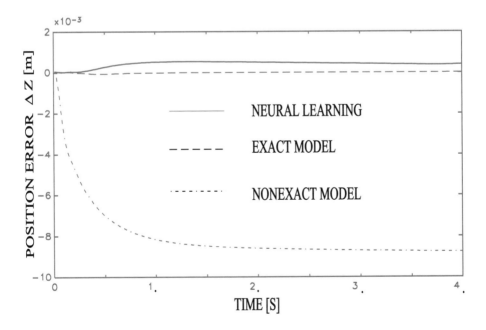

Figure 6.18.   Non-learning and learning impedance control class II - external error ΔZ

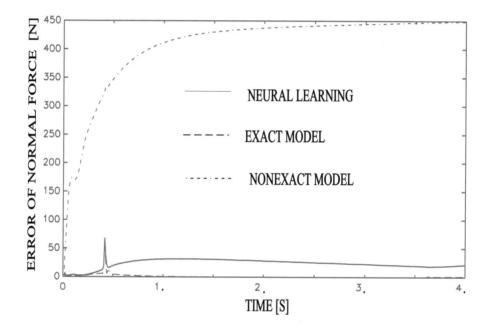

Figure 6.19.   Non-learning and learning impedance control class II - error of normal force

Chapter 7

SYNTHESIS OF COMPREHENSIVE CONNECTIONIST CONTROL ALGORITHMS FOR ROBOT CONTACT TASKS

1.   Introduction

The previously defined connectionist learning control laws for robotic contact tasks did not particularly deal with the problem of environment models and environment uncertainties. The main intention was to compensate for the system uncertainties by connectionist structures in a global fashion, without separately considering the problems of environment modelling, identification, and uncertainties. However, it is well known that most manipulation robots, especially those in industrial practice, are required to operate in uncertain and variable environments. Thus, the characteristics of the environment can be assumed to be unknown and to change significantly, depending on the given task. It is clear that without adequate knowledge about the environment dynamics it is even not possible to determine consistent values of nominal trajectory and force, as well as nominal control, not to mention achieving asymptotic stability. The previous research works in most cases, however, did not consider these uncertainties and nonlinearities of the environment and, thus, the above approaches are limited to specified working conditions, which satisfy only certain assumptions. Namely, when considering specific contact tasks, simplifications in the modelling of robot and environment dynamics are introduced in almost all control approaches, in order to obtain simpler control algorithms. Hence, the uncertain nonlinearities and other characteristics of environment models still remain a critical issue in robotic contact task research.

As a result of the mentioned facts, efficient compliance control algorithms, and especially learning compliance control laws, must include new comprehensive learning features,which are necessary for active compensation of the environment uncertainties and for the determination of optimal control parameters. In this case, algorithms that identify the type of environment models on-line, could significantly improve the performance of contact task control schemes. As one solution, off-line identification of environment parameters based on experimental measurements [300] may also result in good system performance with approximate modelling of a robot dynamic environment that would be sufficiently exact. However, in the case of nonlinear complex models of environment or uncertain structure of the environment model, the conventional parameter identification method is not a solution to the compliance control synthesis.

Hence, in the case of robotic system uncertainties and uncertain parameters and structure of the environment model, some intelligent techniques (fuzzy logic, neural networks, genetic algorithms) can be efficiently applied [352], [46],[339],[176] for robot control and dynamic environment identification in compliance control tasks. In [352], a direct method of environment parameter identification, using recurrent neural networks with terminal attractors for space robotic applications, is proposed. The results indicate good performance in learning and generalization processes. A very interesting approach is the use of intelligent techniques for dynamic environment classification instead of using intelligent techniques for parameter identification. Cha and coworkers [46] use indirect method with neural networks for telerobotic purposes to classify dynamic environments, and fuzzy logic to select force reflection gain based on estimated characteristics of the environment. A combination of different intelligent paradigms (fuzzy + neuro), a special algorithm for the approach, contact and force control of robot manipulators in the presence of a unknown environment is proposed in [176]. In this case, the robot manipulator controller, which approaches, contacts, and applies force onto the environment, is designed using fuzzy logic in order to realize human-like control and then modelled as a neural network to adjust the membership functions and rules to achieve the desired contact force control.

The main idea in the proposed new, comprehensive intelligent control strategy for robotic contact tasks will be the inclusion of connectionist structures with broader senses of application [170], [171]. Beside the "black-box" and hybrid learning of robot and environment uncertainties, the new role of connectionist structures is related to sufficiently exact classification of dynamic environments in order to achieve highly reliable behaviour of the whole robot-environment system. This classification is necessary in order to calculate the control parameters of learning control algorithms that achieve the best performance of the robotic systems. This approach can be considered as a comprehensive connectionist control approach based on extension and generalization of the approach developed and presented in the previous chapter. Because of a new role for connectionist structures, the main feature of the proposed hybrid learning control algorithms will be the integration of two connectionist structures instead of one (as in the previous case) into some typical non-learning control laws (control laws based on the stabilization of robot motion and interaction force and impedance control). The first neural network (neural classifier) will be capable of performing the classification of unknown characteristics of the environment needed to select the appropriate control parameters of the basic non-learning compliance control algorithm. In this way, the learning compliance control algorithm will significantly reduce the influence of robotic system uncertainties. It is important to notice that the neural classifier's objective is to classify the model profile and parameters of the environment in an on-line manner. The classification capability of the neural classifier will be realized by an efficient off-line training process. This off-line training process is defined by a single execution of the same robotic contact tasks but with different dynamic environments and different features of these environments. The classification knowledge of the first neural network is the base for excellent generalization in the on-line procedure through the process of pattern association. As classification tools, two types of connectionist structures will be presented: multi-layer perceptron and wavelet network. The wavelet network classifier is chosen in order to enhance the pattern recognition properties in comparison with the pure multilayer perceptron approach. The wavelet preprocessing includes the necessary feature extraction capability together with classification capability of multilayer perceptrons. On the other hand, the wavelet transformation is very convenient for representing nonstationary signals with brief, high-frequency components as are the signals from the robotic force sensors. In on-line operational mode, after the classification process, the basic learning control algorithm utilizes the information about the type of environment from the first neural network to determine control parameters and the parameters of the environment model. The determination of these parameters can be achieved by values corresponding to the appropriate type of particular environment that is defined in advance, or they can be obtained by the process of linear interpolation. The second proposed neural network (neural compensator) will play the general role of a robust learning controller needed to compensate for uncertainties of the dynamic model of the manipulation robot in contact with a dynamic environment. Hence, the main objective of this approach is to employ coordination of the well-trained neural classifier (first network), together with the neural compensator (second network), to significantly improve the robot's performance in contact tasks when an uncertain environment is involved. Using similar ideas and principles, Kiguchi and Fukuda [180], [181] proposed the fuzzy-neuro control robot force method. The proposed force controller consists of a neural network, which classifies the unknown environment on-line, several fuzzy-neural force controllers suitably designed to apply the initially expected environments, and a fuzzy controller selector, which selects suitable fuzzy-neuro controllers with a proper ratio, according to the properties of the environment. However, it cannot be expected to use human expert knowledge for designing the fuzzy-neural controller in all cases. Hence, the connectionist controller would be a most effective controller for robotic contact tasks. In the last paper [181], GA is applied in combination with neuro-fuzzy force controllers for robotic contact tasks. In this case, multiple genetic neuro-fuzzy force controllers are suitable combined with a proper rate in accordance with the unknown dynamics of the environment. In order to carry out the proposed force control method, several kinds of neuro-fuzzy controllers, designed for different kinds of environments are prepared. The optimal combination rate of the prepared neuro-fuzzy force controllers according to the environment dynamics is defined on-line by a neural network that is off-line trained using genetic algorithms. More exactly, optimal weights of this Controller Combining Neural Network are defined by using GA to output proper weight values for each neuro-fuzzy force controller in accordance with the dynamic property of the environment.

2.   Factors Affecting Task Performance and Stability in Robotic Compliance Control

In order to emphasize the importance of the connectionist learning approach and comprehensive control strategy we shall analyze the factors affecting task performance and stability in compliance control algorithms. As typical compliance control algorithms we consider in particular special control algorithms for stabilizing position and force based on the quality of transient processes [356], together with impedance control algorithms. The main idea of using neural networks for learning a system's uncertainties and for the classification of unknown robot dynamic environments can also be efficiently applied to other types of robot contact control algorithms. This analysis is based on using a dynamic model of the robot interacting with the environment, ma odel of the working environment and the definition of the learning control task defined in Chapter 6. As a first example we consider the control algorithm based on stabilization of the robot motion with a preset quality of transient responses, and it has the following form [356]:

$$\tau = H(q)[\ddot{q}_d - KP\eta - KD\dot{\eta}] + h(q, \dot{q}) + J^T(q)F \qquad (7.1)$$

The family of desired transient responses is specified by the vector differential equation:

$$\ddot{\eta} = -KP\eta - KD\dot{\eta} \qquad (7.2)$$

$$\eta(t) = q(t) - q_d(t) \qquad (7.3)$$

where $KP\epsilon R^{n\times n}$ is the diagonal matrix of position feedback gains; $KD\epsilon R^{n\times n}$ is the diagonal matrix of velocity feedback gains. The right-hand side of equation (7.2), i.e., the PD-regulator, is chosen so that the system defined by (7.2) is asymptotically stable as a whole. The values of the matrices KP and KD can be selected according to the algebraic stability conditions.

The proposed control law represents a version of the well-known computed torque method, including the force term, which uses a dynamic robot model and the available on-line information from the position, velocity and force sensors. In this case, the robot dynamics model (more, the uncertainties of the matrix $H(q)$ and vector $H(q, \dot{q})$) has explicit influence on the performance of the contact control algorithm. Another important characteristic of this control algorithm is that the model of robot environment does not have any influence on its performance . Hence, the influence of different robot environments is expressed through the different values of initial force at the robot tip, i.e., through the different parameters of the environment model. In the initial contact, there are different values of initial force for various robot environments, while all the other model and control parameters are equal. These different initial conditions cause different force transient responses. Also, the inclusion of noise from the robot sensors cause different force steady-state responses for various robot environments. Hence, if the control aim is to achieve the same quality of force steady-state responses for different environments, the same force performance can be achieved only with different values of PD gains. In this case, learning of the matrices of robot dynamic models by the neural network and the determination of PD local gains based on neural classification of the robot environment, will significantly influence the task performance.

As a second example we consider a control algorithm based on stabilization of the interaction force with the preset quality of transient responses, which has the following form [356]:

$$
\begin{aligned}
\tau \;=\; & H(q)M^{-1}(q)[-L(q, \dot{q}) + S^T(q)F] + h(q, \dot{q}) + \\
& J^T(q)\{F_d - \int_{t_0}^{t}[KFP\mu(\omega) + KFI\int_{t_0}^{t}\mu(\omega)dt]d\omega\} \quad (7.4)
\end{aligned}
$$

where $\mu(t) = F(t) - F_d(t)$; $KFP\epsilon R^{n \times n}$ is the matrix of proportional force feedback gains and $KFI\epsilon R^{n \times n}$ is the matrix of integral force feedback gains. Here, it has been assumed that the interaction force in the transient process should behave according to the following differential equation:

$$
\dot{\mu}(t) = Q(\mu) \quad (7.5)
$$

$$
Q(\mu) = -KFP\mu - KFI\int_{t_0}^{t}\mu dt \quad (7.6)
$$

whereby the PI-force regulator (continuous vector function Q) is chosen such that the system defined by (7.5) is asymptotically stable as a whole.

In this case, it is clear that the uncertainties of the robot dynamics model have explicit influence on the performance of the contact control algorithm. However, the environment dynamics model has explicit influence on the performance of the contact control algorithm, also influencing the PI force local gains. It is clear that if the knowledge of the environment model (parameters of the matrices $M(q), L(q, \dot{q}), S(q))$ is not accurate enough, it is not possible to determine the nominal contact force $F_p(t)$. Moreover, an inexact model of environment dynamics can significantly influence the contact task performance. Hence, if the aim of the control is to obtain the same quality of force steady-state processes for different environments, the same force performance can be achieved only by determining the environment model profile and/or identifying parameters of the robot environment model, and using equal fixed PI force local gains. The unknown model profile and/or parameters of the robot environment model have a greater significance for the system performance in comparison with the PI force local gains. In this case, the model profile of the environment and the parameters of the robot environment model based on classification of the robot environment, affect the task performance. The uncertainties of the robot dynamic model are another factor affecting the task performance.

As a third example of compliance control laws, the impedance control algorithm in the case of environment model can be given as

$$\tau = H(q)[\ddot{q}_0 + M^{-1}(-B\dot{\epsilon} - K\epsilon + F - F_0)] + h(q, \dot{q}) + J^T(q)F \quad (7.7)$$

where $\epsilon = q - q_0$ is the position error.

In this case, it is evident that in addition to the uncertainties of the robot dynamic model, the uncertainties of environment impedance models (matrices M, B, K) have a great influence on the performance of the basic compliance control algorithm. The aim of neural network learning and classification is the determination of the exact parameters of the robot dynamic model, compensation of model uncertainties, and classification of the environment impedance model profile and/or impedance parameters.

3.   Comprehensive Connectionist Control Algorithm Based on Learning and Classification for Compliance Robotic Tasks

The main objective in the compliance control strategy is the application of connectionist learning structures as part of the stabilizing control algorithm or impedance control algorithm presented in Chapter 6. The role of the proposed connectionist structure is to compensate for the possible uncertainties and differences between the real robot dynamics and the assumed dynamics defined by the user in the process of control synthesis. However, the proposed learning control algorithms do not work in a satisfactory way if there is no sufficiently accurate information about the type of robot environment model and the parameters of the environment model. Hence, in order to enhance connectionist learning of the general robot-environment model, a new connectionist control strategy is proposed, based on learning and classification properties. The main idea of the proposed strategy is the use of the neural network approach as a classification and learning tool. The connectionist classifier will be trained through an off-line process to recognize robot dynamic environments and will be used to sufficiently exactly classify robot dynamic environments in the on-line regime by generalization capabilities. Its objective is to classify the model profile and environment parameters in an on-line manner. Hence, the application of the connectionist approach to this type of problem is divided into two phases: <u>first</u>, related to the acquisition process and off-line training of the proposed neural network and, <u>second</u>, association phase, where on-line control algorithms, based on excellent generalization properties of the neural network classifier and learning properties of the neural compenstor, should ensure the necessary quality of the system performance.

3.1   Acquisition process for classification - the first phase

In the acquisition process of the first phase, based on the repeated realization of the proposed contact control algorithms using a previously chosen set of different working environments, force data from force sensors are observed and stored in specific data files. For each robot environment and for the same chosen contact control algorithm, the values of normal force and error of normal force are measured, calculated and stored as special input patterns for training the neural classifier. Generally, six contact force components can be collected during the task realization, but the attention will be focused only on the normal force, as one of the most interesting components, which is sufficient to classify the unknown environment characteristics. It is assumed that the normal force component can be obtained from the force sensor, because normal and tangential directions of force components are defined when considering machining operations. On the other hand, the acquisition process has to be accomplished using various robot environments, starting with an environment with a low level of system characteristics (for example, with a low-level environment stiffness) and ending with the environment having a high level of system characteristics (with a high-level environment stiffness). As with other important characteristics in the acquisition process, different model profiles of the environment are used, based on additional damping and stiffness members, which are added to the basic general impedance model. It is important to notice that the main idea is the classification of the environment type, not only environment parameter identification. This approach represents a good foundation for encompassing a wide range of unknown robot environment characteristics.

3.1.1   Pure neural network classifier

Connectionist classifiers represent part of the intelligence pattern recognition methods for classifying technical signals. Many classifiers are based on computing the distances, in an appropriate metric, between the input pattern vector y and the representatives $x_1, ..., x_m$ of M patterns of classes $\omega_1, ..., \omega_m$. If the input pattern y is "closer" to x_j than to any of the other representatives, it is said to belong to the class ω_j. The neural classifiers extend this concept by training (learning) a dynamic system $z(t) = Wz(t)$ that has the representatives $\{x_j\}$ as test points. If y belongs to the domain of attraction of x_j, as the dynamical system is run with initial data y, one has that $z(t) \to x_j$ as $t \to \infty$.

During the extensive off-line training process, the *neural classifier* receives a set of input-output patterns, where input variables form a previously collected set of force data (in each learning iteration, the normal force $F_n(t)$ and the error of normal force $\Delta F_n(t)$ in the time instants $(t), (t-1), (t-2)$ and $(t-3)$ during the force transient process). The choice of error of normal force in the preceding time instants is determined because of the fact that the nonlinear mapping depends on the previous inputs and outputs of the system. The neural classifier has as desired output a value between zero and the value defined by the environment profile model (the whole range between 0 and 1) which exactly defines the type or training robot environment (different model parameters) and type of the environment model profile. The aim of connectionist training is that the real output of a neural network for given inputs can be determined exactly or very close to the desired output value, for appropriate training robot environment and environment model profile.

As an example that is applied in simulation experiments, the neural network training will be accomplished with five different working environments and five different environment models. The environment profile models are shown in Table 7.1.

Table 7.1.   Environment model profiles

Model profile
$F = M'\Delta\ddot{x} + B'\Delta\dot{x} + K'\Delta x$
$F = M'\Delta\ddot{x} + B'\Delta\dot{x} + K'\Delta x + K_1\Delta x^3$
$F = M'\Delta\ddot{x} + B'\Delta\dot{x} + K'\Delta x + B_1\Delta\dot{x}^3$
$F = M'\Delta\ddot{x} + B'\Delta\dot{x} + K'\Delta x + K_1\Delta x^3 + K_2\Delta x^2$
$F = M'\Delta\ddot{x} + B'\Delta\dot{x} + K'\Delta x + B_1\Delta\dot{x}^3 + B_2\Delta\dot{x}^2$

where B_1, B_2, K_1, K_2 are the adjoint matrices for damping and stiffness members.

The input variables and target outputs for different environments are shown in Table 7.2.

Table 7.2. Inputs and target outputs of neural classifier

Input data for classifier	Classifier outputs
$F_n(t)$	Styrofoam 0.00
$\Delta F_n(t)$	Silicon 0.25
$\Delta F_n(t-1)$	Rubber 0.50
$\Delta F_n(t-2)$	Plastic 0.75
$\Delta F_n(t-3)$	Steel 1.00

It will be generally assumed that the training examples represent specific and commonly used environment profiles. Hence, the success of the classification will be determined by the "richness" of the training examples.

A neural network classifier based a on four-layer perceptron is chosen for the purpose of classification due to good generalization properties. Apart from the choice of perceptron topology, it is very important to choose an efficient learning algorithm for the off-line training process, to ensure best convergence of the learning process. Hence, we consider the learning algorithms for adjusting the network weights based on the application of the RLS method with gradient approximation [165]. The use of these methods with time-varying learning rate yields benefits for learning speed and generalization compared to those available with the standard back propagation algorithm.

The main relations in the process of training for forward pass in a 4-layer network are described by the following expressions:

$$s^2(k) = W^{12}(k)^T i^1(k) \tag{7.8}$$
$$o_a^2(k) = 1/(1 + exp(-s_a^2(k)))a = 1, ..., L_1 \quad o_0^2(k) = 1 \tag{7.9}$$
$$s^3(k) = W^{23}(k)^T o^2(k) \tag{7.10}$$
$$o_b^3(k) = 1/(1 + exp(-s_b^3(k))) \quad b = 1, ..., L_2 \quad o_0^3(k) = 1 \tag{7.11}$$
$$s^4(k) = W^{34}(k)^T o^3(k) \tag{7.12}$$
$$y_c(k) = s_c^4(k) \quad c = 1. \tag{7.13}$$

where $s^2(k)$, $s^3(k)$, $s^4(k)$ are the output vectors for linear parts of the layers at a time instant k; $o^2(k)$, $o^3(k)$ are the output vectors of the hidden layers; $W^{12} = [w_{m \times L1}^{12}]$, $W^{23} = [w_{L1+1 \times L2}^{23}]$, $W^{34} = [w_{L2+1 \times n}^{34}]$ are the weighting factors of the layers; $i^1(k)$ are the inputs to the network (force data $m = 5+$ bias member $= 6$); $y(k)$ is the network output at the time instant k.

The general idea is that multilayer perceptrons can be observed as a set of sequential linear decomposed subsystems, which are connected by nonlinear connections. As is known, recursive estimators for linear deterministic systems show faster convergence properties than gradient estimators (BP algorithm). Hence, the aim of estimation is to define optimal values for the matrices W^{12}, W^{23}, W^{34}. In the application of this method to specify desired values for the linear parts of the layers $s^{id}(i = 2, 3)$, use is made of gradient approximation. The basic equations describing the proposed learning rules based on the RLS gradient approximation method are given as [165]:

$$
\begin{aligned}
for \quad h \ &= \ 4 \\
s_c^{hd}(k) \ &= \ y_c^d(k) \quad c = 1.
\end{aligned}
\tag{7.14}
$$

$$
\begin{aligned}
for \quad h \ &= \ 3, 2 \quad g = 4, 3 \\
s_t^{hd}(k) \ &= \ s_t^h(k) + \alpha_t \delta^g(k)^T W^{hg}(k)^T o^h(k)[1 - o_t^h(k)] \\
&\qquad t = 1, ..., L_2 \quad or \quad t = 1, ..., L_1
\end{aligned}
\tag{7.15}
$$

$$
\begin{aligned}
for \quad h \ &= \ 3, 2, 1 \\
C^h(k) \ &= \ C^h(k-1) - \frac{C^h(k-1)o^h(k)o^h(k)^T C^h(k-1)}{1 + o^h(k)^T C^h(k-1)o^h(k)}
\end{aligned}
\tag{7.16}
$$

$$
\begin{aligned}
for \quad h \ &= \ 3, 2, 1 \quad g = 4, 3, 2 \\
W^{hg}(k) \ &= \ W^{hg}(k-1) + C^h(k)o^h(k)[s^{gd} - W^{hg}(k-1)^T o^h(k)]^T
\end{aligned}
\tag{7.17}
$$

$$
\begin{aligned}
for \quad h \ &= \ 4, 3 \quad g = 3, 2 \\
\delta_t^h(k) \ &= \ s_t^{hd}(k) - \sum_j w_{tj}^{gh}(k)o_j^g(k) \\
&\qquad t = 1, .., n \quad j = 1, .., L_2 + 1 \quad or \\
&\qquad t = 1, .., L_2 \quad j = 1, .., L_1 + 1
\end{aligned}
\tag{7.18}
$$

where α_t is the appropriate learning rate; C^h is the appropriate covariance matrix; y_c^d is the desired output of the network.

3.1.2 Wavelet network classifier

In order to enhance the connectionist learning of a general robot-environment model, wavelet network classification of the robot dynamic environment is proposed as another solution. Namely, in the pattern recognition tasks it is important to use *feature extraction* from the occurring patterns, which is the conversion of patterns to features that are regarded as a condensed representation, ideally containing all important information. Hence, the proposal of the new classifier represents the combination of the advantages of a multilayer perceptron with the appropriate feature-extraction methods [329], by designing a classification procedure which consists of two parts: a) small numbers of features are calculated by wavelet transformation from high-dimensional input patterns (signals from force sensors) and b) the calculated features are regarded as inputs to the simple multilayer perceptron, which is used as final classifier.

To summarize the essential components of classifying procedure we will describe the process of feature extraction by wavelet transformation [51]. The signals from the robot force sensors are frequently characterized by nonstationary time behaviour. For this type of signals, time-frequency representation is highly desirable in order to derive important features. From a variety of different approaches to the theory of wavelets and signal processing [118], [51], [282] we will present one specific method for feature extraction based on the linear one-dimensional wavelet transformation $w(a, \tau)$. In mathematical terms, the wavelet transformation is expressed as the inner product of a signal $s(t)$ with a basic wavelet function $h_a(t)$:

$$w(a, \tau) = \langle s; h_a \rangle = \frac{1}{\sqrt{a}} \int s(t) h_a^* (\frac{t - \tau}{a}) dt \qquad (7.19)$$

where a is the scale parameter, τ is the shift parameter of the wavelet function; * denotes the complex conjugate. The weighting factor $\frac{1}{\sqrt{a}}$ normalizes the wavelet function for all values of a to constant energy.

The wavelet transformation $w(a, \tau)$ can be considered as the correlation function between the signal $s(t)$ and the basic wavelet h_a. Typically, basic wavelets are modulated windows, i.e., they compose product of two functions. There are various basic wavelet families such as Mexican hat wavelet, Meyer wavelet and Morlet wavelet. As a typical example we can use the complex Morlet wavelet:

$$h_{Ma} = \frac{1}{\sqrt{a}} exp[-0.5(\frac{t}{a})^2 + j2\pi f_0 \frac{t}{a}] \qquad (7.20)$$

Analysis by wavelet transformations can be generally considered as a filter bank composed of bandpass filters with the bandwidths proportional to frequency. In this case, the filter bank is characterized by the particular basic wavelet function and the parameter f_0. Depending on the scale factor a, the time course of the wavelet transformation can be changed. If a is large, the wavelet is a dilated low-frequency function, while for small values of a, the wavelet is constricted, corresponding to a high-frequency function. Important features of extraction that are suited for distinguishing different patterns can be defined as particular values of $w(a, \tau)$, at specific times τ_k, and scale factors a_k, respectively, for the frequency values $f_k = \frac{f_0}{a_k}$.

The concept of the wavelet network [329] for classification purposes is represented by an expanded perceptron for decision purposes together with the so-called wavelet nodes as preprocessing units for feature extraction. It is important to notice that during the learning phase, the wavelet network not only learns the appropriate decision functions and complex decision regions defined by the weight coefficients, but also searches for those parameter spaces that are suited for a reliable categorization of the input force signals. In Figure 7.1 is shown a basic version of the wavelet network structure.

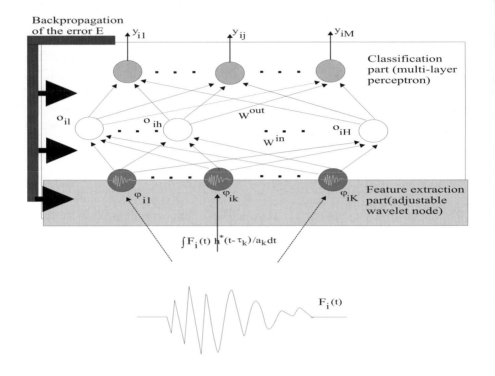

Figure 7.1. Wavelet network classifier

The wavelet nodes, which are adjusted during the learning phase, are a modified version $h(t - \tau_k/a_k)$ of the basic wavelet $h(t)$. The nodes are described by a time shift parameter τ_k, and a scale parameter a_k, which is inversely related to the node frequency (τ_k and a_k are the parameters of the wavelet transformation). The input to the wavelet network, i.e., the output of the wavelet node ϕ_{ik}, represents the inner product of the node h_k and the signal from the robotic force sensor F_i (the inxx i denotes the signal samples, $i = 1, ..., N$):

$$\phi_{ik} = \langle h_k; F_i \rangle = \int_t h * \left(\frac{t - \tau_k}{a_k}\right) F_i(t) dt$$

The upper part of the wavelet network is represented by the topology of the 3-layer perceptron, which bases its classification decision on the wavelet node output. Training of the perceptron can be achieved by using minimization of the least-squares error ($M = 1$ in this case):

$$E = \sum_{i=1}^{N}(d_i - y_i)^2 = min \qquad (7.21)$$

where d_i is the desired output vector that belongs to an appropriate class of training patterns. It is important to notice that not only the weights of the network are thus adjusted, but also the parameters of the wavelet nodes. The parameters of the wavelet transformation, the scale parameter a_k and the shift parameter τ_k, depend on the basic wavelet chosen and they are determined using, for example, the well-known back-propagation algorithm:

$$\frac{\partial E}{\partial \tau_k} = -\sum_{i=1}^{N}\sum_{j} \delta_{ij} w_{kj} \frac{\partial \phi_{ik}}{\partial \tau_k} \qquad (7.22)$$

$$\frac{\partial E}{\partial a_k} = -\sum_{i=1}^{N}\sum_{j} \delta_{ij} w_{kj} \frac{\partial \phi_{ik}}{\partial a_k} \qquad (7.23)$$

$$\delta_{ij} = (d_i - y_i)y_i(1 - y_i) \qquad for \qquad output \qquad layer \qquad (7.24)$$

$$\delta_{ij} = O_j^{hid}(1 - O_j)^{hid})w_j^{out}(d_i - y_i)y_i(1 - y_i) \quad for \quad hidden \quad layer$$
$$(j = 1, ..., H) \qquad (7.25)$$

$$O_j^{hid} = f(\sum_{k} w_{kj}^{in}\phi_{ik} + b_j) \qquad (7.26)$$

where f is a sigmoid or other activation function; b_j is a bias member.

In our example, a complex Morlet wavelet is applied as basic wavelet. Hence, the wavelet node can be calculated in the following way:

$$\phi_{ik} =| \int_t h * (\frac{t - \tau_k}{a_k} F_i(t)dt |= \sqrt{(\int_t Ficos\ dt)^2 + (\int_t Fisin\ dt)^2} \quad (7.27)$$

where

$$Ficos = F_i(t) \cos(\omega_k \frac{t - \tau_k}{a_k}) \exp(-0.5(\frac{t - \tau_k}{a_k})^2) \qquad (7.28)$$

$$Fisin = F_i(t) \sin(\omega_k \frac{t - \tau_k}{a_k}) \exp(-0.5(\frac{t - \tau_k}{a_k})^2) \qquad (7.29)$$

where the frequency parameter $\omega_k = 2\pi f_k$ which, in contrast to the assumption of a fixed value f_0, results in greater flexibility.

Using summation for integration in discrete processing, and according to the formulae 7.22 - 7.27, the adjustment of parameters of the wavelet transformation is described by the following equations:

$$\frac{\partial E}{\partial \tau_k} = \sum_{i=1}^{N} \sum_{j} \delta_{ij} w_{kj} \frac{1}{\phi_{ik}} \sum_{m=1}^{T_m} F_i(t_m) \exp(-0.5(\frac{t_m - \tau_k}{a_k})^2) \frac{1}{a_k}$$

$$(\int_t Ficos \; dt[\omega_k \sin(\omega_k \frac{t_m - \tau_k}{a_k}) + \frac{t_m - \tau_k}{a_k} \cos(\omega_k(\frac{t_m - \tau_k}{a_k}))]$$

$$+ \int_t Fisin \; dt[-\omega_k \cos(\omega k \frac{t_m - \tau_k}{a_k}) + \frac{t_m - \tau_k}{a_k} \sin(\omega_k(\frac{t_m - \tau_k}{a_k}))])$$

$$(7.30)$$

$$\frac{\partial E}{\partial a_k} = \frac{t - \tau_k}{a_k} \frac{\partial E}{\partial \tau_k} \tag{7.31}$$

$$\frac{\partial E}{\partial \omega_k} = \sum_{i=1}^{N} \sum_{j} \delta_{ij} w_{kj} \frac{1}{\phi_{ik}} \sum_{m=1}^{T_m} F_i(t_m) \exp(-0.5(\frac{t_m - \tau_k}{a_k})^2) \frac{t_m - \tau_k}{a_k}$$

$$[-\int_t Ficos \; dt \sin(\omega_k \frac{t_m - \tau_k}{a_k}) + \int_t Fisin \; dt \cos(\omega_k \frac{t_m - \tau_k}{a_k})]$$

$$(7.32)$$

where T_m is the number of signal sampling points.

After the acquisition process, during the extensive off-line training process, in the same way as the perceptron, the wavelet network receives a set of input-output patterns, where input variables are the wavelet transformation of force signals, while the desired network output has a value between zero and unity, which exactly defines the type of training of the robot environment.

3.2   On-line compliance control algorithms for contact tasks with environment classification - the second phase

Based on the first phase, related to the acquisition process and off-line training process of the neural classifier, it is possible to determine the whole structure of compliance control algorithms, including the neural compensator and neural classifier. After the off-line training process with different working environments and different environment model profiles, the neural classifier is included in the on-line version of the control algorithm to produce some value between 0 and 1 at the network output. Based on this value, through the process of linear interpolation, the environment parameters $M(q), L(q, \dot{q})$, along with the environment model structure, are effectively determined. This interpolation process is driven by the stored parameters of the dynamic models of different chosen environments and different chosen environment model structures.

In the first example considered, the following stiffness model of robot environment is chosen for control algorithm based on the stabilization of the robot motion with a preset quality of transient process:

$$F = K'(x - x_0) \tag{7.33}$$

After the off-line training process with different working environments (different environment stiffness), the neural classifier with fixed weighting factors is included in the on-line version of control algorithm (7.1) to produce some value y at the network output being between 0 and 1, based on the on-line force inputs defined in the previous section:

$$\tau = \hat{H}(q)[\ddot{q}_d - \hat{KP}\eta - \hat{KD}\dot{\eta}] + \hat{h}(q, \dot{q}) + J^T(q)F + P^{NN} \tag{7.34}$$

$$\hat{KP} = f_{kp}(y) \tag{7.35}$$

$$\hat{Kd} = f_{kd}(y) \tag{7.36}$$

where f_{kp}, f_{kd} are the linear interpolation functions for the position and velocity feedback gains; y is the output of the neural classifier.

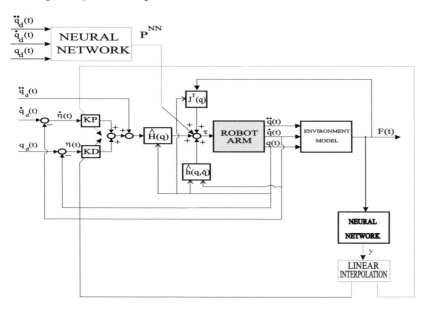

Figure 7.2. Scheme of learning control law stabilizing robot motion with neural classifier

If we adopt the same force steady-state process for all different robot environments as performance criterion, then, using algebraic stability conditions, we can a priori choose a set of PD local gains for the previously defined set of known robot environments (in our case there are five different environments) that will satisfy this requirement. Hence, in the case of unknown environment type, the information from the neural classifier output can be efficiently utilized for calculation of the necessary PD local gains by linear interpolation procedures. It is also assumed that the neural network output for the given environment varies with in a small range. In this way, the local PD gains are relatively fixed during the operations. They are chosen for presetting stability conditions for each environment type. Figure 7.2 shows the overall structure of the proposed algorithm.

In the second example, for a control algorithm based on the stabilization of interaction force with a preset quality of transient process, the general impedance model of robot environment is chosen:

$$F = M'\Delta\ddot{x} + B'\Delta\dot{x} + K'\Delta x \tag{7.37}$$

Hence, after the off-line training process, the on-line version of the compliance control algorithm with the neural classifier, having fixed weighting factors based on the on-line force and force errors inputs for a specified environment model, is given by the following relations:

$$
\begin{aligned}
\tau \;=\; & -H(q)\hat{M'}^{-1}(q)[\hat{B'}\dot{q} + \hat{K'}q] + h(q,\dot{q}) + \\
& (J^T(q) - H(q)\hat{M'}^{-1})\{F_d - \int_{t_0}^{t}[KFP\mu(\omega) + KFI\int_{t_0}^{t}\mu(\omega)dt]d\omega\} \\
& +P^{NN}
\end{aligned}
\tag{7.38}
$$

$$\hat{M'} = f_{M'}(y) \tag{7.39}$$

$$\hat{B'} = f_{B'}(y) \tag{7.40}$$

$$\hat{K'} = f_{K'}(y) \tag{7.41}$$

where $f_{M'}, f_{B'}, f_{K'}$ are the linear interpolation functions for the parameters of the matrices M', B', K'.

According to a similar principle, the same condition for control laws and all different robot environments assumes the same local PI force gains. In our case, the parameters of dynamic models of the different chosen environments M', B', K' are stored as the information needed to calculate the basic control algorithm. In the case of an unknown environment, the information from the neural classifier output can be efficiently utilized to calculate the necessary environment parameters M', B', K' by linear interpolation procedures. Figure 7.3 shows the overall structure of the proposed algorithm.

It is important to notice that the previous equations are determined according to the appropriate environment model profile (stiffness and general impedance

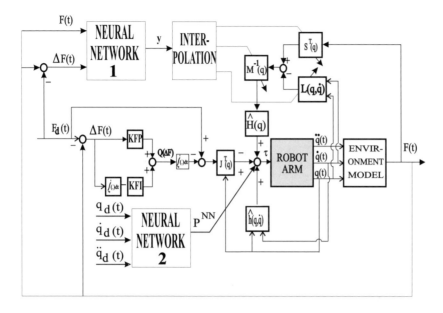

Figure 7.3. Scheme of control law stabilizing interaction force with neural classifier

model). For other types of model profile, the control equation and linear interpolation procedure can be determined in a similar way.

In the same way, the wavelet network classifier can be included in a comprehensive compliance learning control law, as is shown in Figure 7.4 for the case of stabilization of the interactive force. The same principle of on-line linear interpolation with neural classifier could be included in comprehensive learning impedance control law.

4.   Case Study

To demonstrate the performance of contact control schemes with the connectionist classifier and compensator, compliance control implementations are simulated using the PUMA 560 robot for the circular writing task in various robotic environments. The robotic end effector exerts a force that is perpendicular to the y-z plane, performing a circular path with a diameter $d=12$[cm]. The task of the robot is to carry out the writing process on the work surface along the prescribed trajectory with a desired contact force $F_N^0 = 5 \; [N]$.

For the first example of a stabilizing motion control algorithm, the stiffness model of the environment is adopted, while in the case of the stabilizing interaction force algorithm, a general model of impedance is chosen. The parameters of the environment model in the form of diagonal members of appropriate ma-

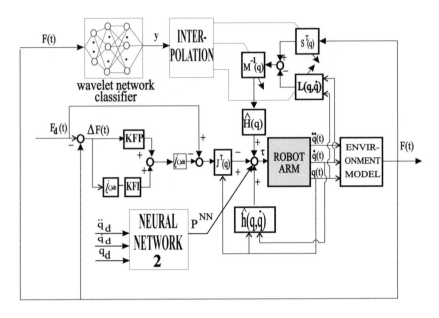

Figure 7.4. Scheme of learning control law with wavelet network classifier

trices for all different chosen environments and for both control algorithms are given in Tables 7.3 - 7.5.

Table 7.3. Stiffness parameters of robot environment models

Environment	K'_{11}	K'_{22}	K'_{33}	K'_{44}	K'_{55}	K'_{66}
Styrofoam	660.	6600.	660.	0.0007	0.0007	0.0007
Silicon	2000.	20000.	2000.	0.002	0.002	0.002
Rubber	3300.	33000	3300.	0.003	0.003	0.003
Plastic	6000.	60000.	6000.	0.006	0.006	0.006
Steel	24000.	240000.	24000.	0.02	0.02	0.02

To investigate the effect of different chosen PD local gains for stabilizing the motion control law and various robot environments, simulation experiments were conducted with two sets of PD local gains. One set of gains is chosen for low stability degree, while the other is chosen for high stability degree. Internal coordinates error and force error in this case are presented in Figures 7.5 - 7.8. The feedback gains for stabilizing control laws with the required quality of position transient response have been chosen in the form of diagonal matrices

Table 7.4. Damping parameters of robot environment models

Environment	B'_{11}	B'_{22}	B'_{33}	B'_{44}	B'_{55}	B'_{66}
Styrofoam	69.	277.	69.	0.007	0.007	0.007
Silicon	210.	839.	210.	0.02	0.02	0.02
Rubber	346.	1385	346.	0.035	0.035	0.035
Plastic	629.	2517.	629.	0.06	0.06	0.06
Steel	2638.	10704.	2638.	0.27	0.27	0.27

Table 7.5. Inertia parameters of robot environment models

Environment	M'_{11}	M'_{22}	M'_{33}	M'_{44}	M'_{55}	M'_{66}
Styrofoam	1.88	6.7	1.88	0.007	0.007	0.007
Silicon	5.7	20.3	5.7	0.02	0.02	0.02
Rubber	9.4	33.5	9.4	0.03	0.03	0.03
Plastic	17.	60.9	17.	0.06	0.06	0.06
Steel	68.15	243.41	68.15	0.2	0.2	0.2

for the first set of gains (low stability degree):

$$KP = diag\{kp^i\}, \quad KD = diag\{kd^i\} \quad kp^i = 1. \quad kd^i = 2. \quad i = 1, ..., n$$
$$(7.42)$$

and for the second set of gains (high stability degree):

$$KP = diag\{kp^i\}, \quad KD = diag\{kd^i\} \quad kp^i = 100. \quad kd^i = 20. \quad i = 1, ..., n$$
$$(7.43)$$

We can observe the dependence of transient processes on the type of robot environment and chosen set of PD gains. In the case of high gain feedback, there are differences only in the beginning of the transient process. In the case where force noise is included, different force steady-state processes for different working environments are observed (Figure 7.9).

Hence, for our reference case, the following performance criterion is chosen: the sum of force error during the task cannot be greater than 11. To achieve this performance criterion, different local PD gains for different environments must be synthesized, based on simulation experiments (Table 7.6).

In the process of neural network training, 500 force training patterns were used (for all 5 different environments with the same control law and different previously chosen PD local gains there are 100 input-output patterns). After

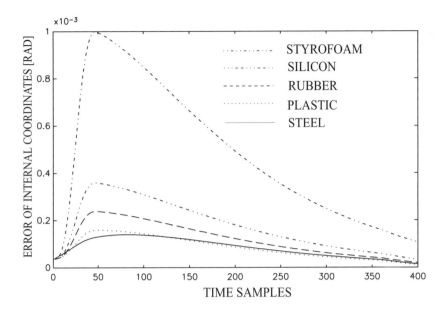

Figure 7.5. Force error - control law stabilizing robot motion - first set of feedback gains

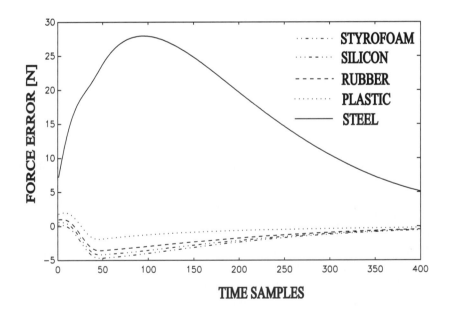

Figure 7.6. Internal error - control law stabilizing robot motion - first set of feedback gains

Table 7.6. PD local gains for satisfaction of performance criterion

Environment	KP	KD
Styrofoam	100.	20.
Silicon	484.	44.
Rubber	900.	60.
Plastic	2209.	94.
Steel	10000.	200.

intensive simulation experiments, the following network topology was chosen: 5 - 62 -41 -1 (numbers of neurons in the network layers). According to the proposed learning rules, initial values of covariance matrix are $C_{init} = 100000I$ and the gradient factor $\alpha_t = 0.02$. The training results are shown in Figures 7.10-7.11, which represent square criterion values during training and the comparison of the desired and real outputs of the network after training.

In the generalization test, the "learned" neural classifier with fixed weighting factors is included in the control algorithm for recognition of the unknown robot environment. In this case, the robot environment with the dominant stiffness K'_{22} is selected. The goal is to achieve the same quality of force steady-state process. The neural classifier. based on input force data, generates the network output having numeric values of 0.62. This information is necessary for calculating by linear interpolation the local PD gains, that can satisfy the desired performance criterion (Figure 7.12). For comparison, an example of the application of non-learning control laws with inexact (user-assumed) information of environment stiffness is given in Figure 7.13. It is clear that in the case when there is no exact information about the robot environment, the quality of performance is very poor. Hence, inclusion of the neural classifier is very important in order to improve the capabilities of control algorithms in working environments with a significant level of uncertainty.

In the second case, for the application of the stabilizing force interaction control algorithm, a performance criterion based on selection of the same force PI gains is chosen. These PI force gains are synthesized using the same system frequencies for all different working environments ($\omega_n = 2Hz$.). The transient processes of internal coordinates error and force error are given in Figures 7.14-7.15. We can notice the influence of different working environments.

In the phase of connectionist training, the following optimal network topology is selected: 6-32-21-1. Using the adopted network topology and the same learning rules and learning parameters, training is accomplished with the stored weighting factors.

After the acquisition process, off-line neural network training was performed for both types of compliance control algorithms. During the training, 2000 force training patterns were used. The following topology for the neural classifier was chosen: 6 - 54 -38 -1. According to the proposed learning rules, the initial values of the covariance matrix is $C_{init} = 100000I$ and the gradient factor $\alpha_t = 0.01$. The uncertainties of the robot dynamics model and dynamic environment model are defined by parametric disturbances and additional white noise.

In a way similar to that in the previous case, the generalization test with unknown environments (the dominant stiffness K'_{22} and the output of the neural classifier 0.70), using the approaches with and without neural classifier, is performed. The second neural network for uncertainty compensation uses the same learning rules and parameters but a different network topology (31-69-37-6). The profile model of the environment, using a general impedance model with additional stiffness members, was adopted. The results are given in Figures 7.16 - 7.17. The conclusions are the same as in the case of stabilizing robot motion control algorithms, but the influence of unknown environment is in this case very significant because of the implicit inclusion of environment parameters in the control law. Hence, the neural classifier significantly improves the system performance. Also, an improvement in learning system uncertainties is acieved through learning epochs is.

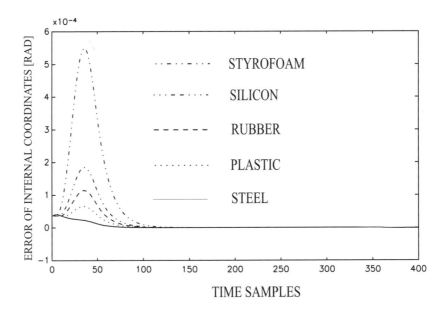

Figure 7.7.   Force error - control law stabilizing robot motion - second set of feedback gains

In the example of a non-learning impedance algorithm, the matrices of target impedance are set according to the values from the paper [169]. The interaction force for different working environments in the acquisition process is shown in Figure 7.18. After the acquisition process, the off-line wavelet network training is performed. The following network topology was chosen: 7 (number of wavelons in the wavelet node) - 42 -24 -1. The uncertainties of the robot dynamics model and dynamic environment model are defined by parametric disturbances and additional white noise. In the generalization test, the "off-line learned" wavelet neural classifier with fixed weighting factors was included in the control algorithm for the recognition of unknown robot environments. The second neural network for uncertainty compensation utilized the same learning rules and parameters but had a different network topology $(31 - 69 - 37 - 6)$. The profile model of the environment, using the general impedance model with additional stiffness members, was adopted. In this case, the robot environment with the dominant stiffness $K = 65000N/m$ was selected. The wavelet neural classifier, based on input force data, generates the appropriate value at the network output. For comparison, an example of the application of impedance learning control laws with and without exact information of environment stiffness after first learning epoch is given in Figure 7.19. It is evident that in the case where there is no exact information about the robot environment, the quality of the position tracking performance is poor.

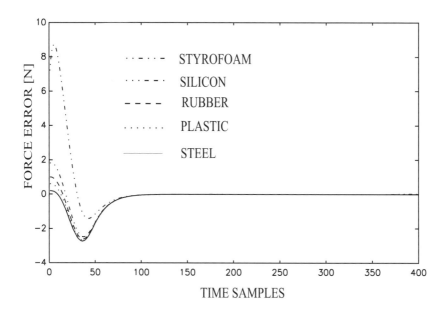

Figure 7.8. Internal error - control law stabilizing robot motion - second set of feedback gains

In this chapter we have presented a new, comprehensive method for selecting the appropriate compliance control parameters for robotic contact tasks. The method classifies the type of environment by using in the first phase acquisition process of force sensor data and off-line training process by multilayer perceptrons of the wavelet networks. The important feature is that the process of pattern association can work in an on-line mode as a part of the selected compliance control algorithm. Simulation experiments showed that a neural classifier together with the neural compensator, can significantly improve robot performance in the contact tasks involving environment uncertainties.

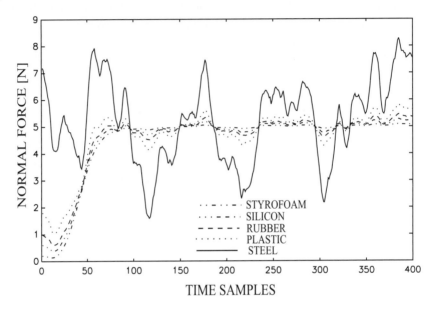

Figure 7.9.   Steady-state process for normal force - control law stabilizing robot motion - second set of feedback gains

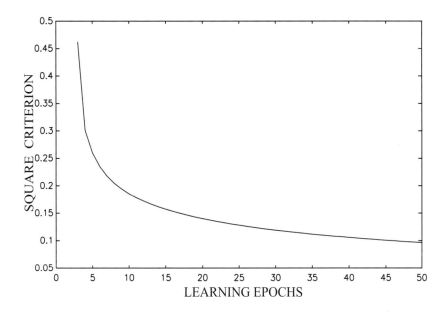

Figure 7.10.   Square criterion during learning epochs

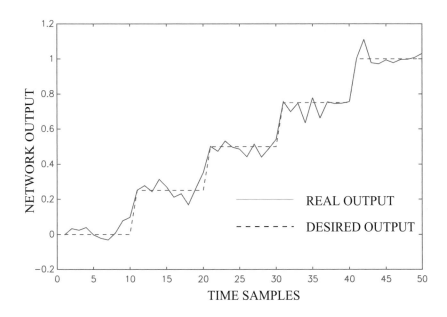

Figure 7.11.   Comparison of desired and real output of network

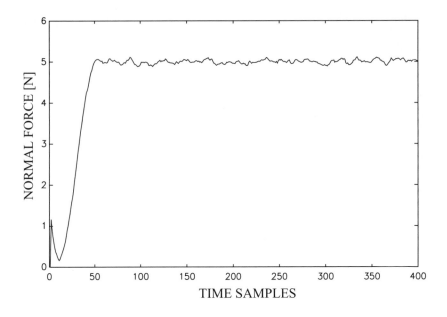

Figure 7.12.   Normal force with neural classifier - stabilizing robot motion

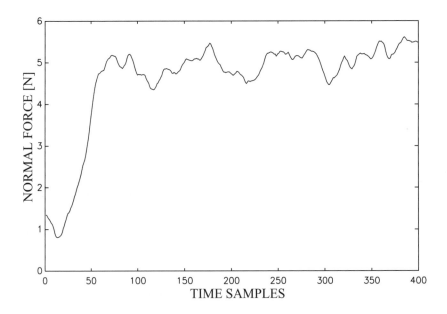

Figure 7.13.   Normal force without neural classifier - stabilizing robot motion

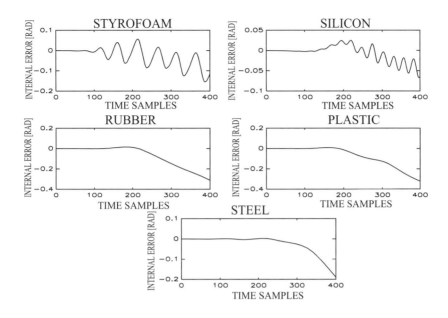

Figure 7.14. Internal error for stabilizing interaction force control algorithm

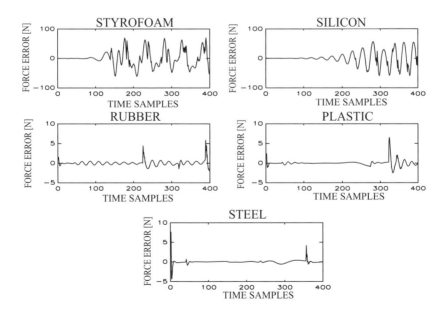

Figure 7.15. Force error for stabilizing interaction force control algorithm

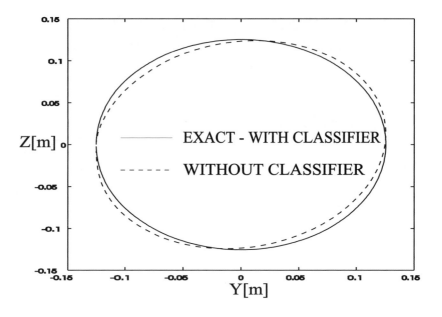

Figure 7.16. Circular tracking - comparison with and without neural classifier

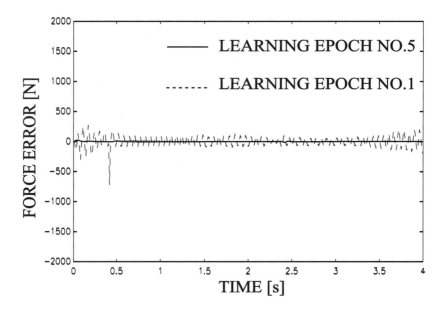

Figure 7.17. Force error - comparison with and without neural classifier

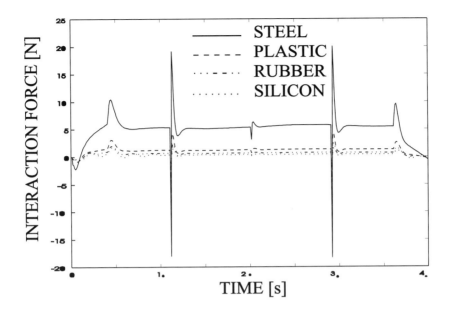

Figure 7.18. Interaction force - non-learning impedance control algorithm

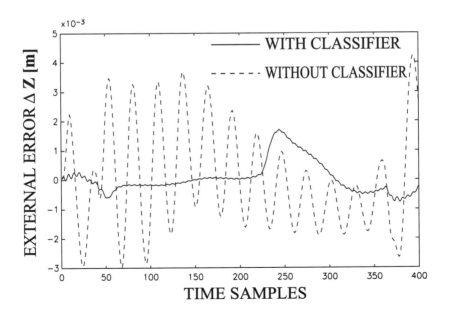

Figure 7.19. Comparison with and without classifier

Chapter 8

EXAMPLES OF INTELLIGENT TECHNIQUES FOR ROBOTIC APPLICATIONS

1. Introduction

The main purpose of this chapter is to present some interesting examples of connectionist approaches together with some supporting intelligent techniques for special robotic applications, to verify the efficiency of contemporary intelligent control methods.

As the first example, connectionist reactive control for a robotic assembly task by soft sensored grippers is presented [34]. In this example, the problem of efficient control in assembly tasks realized by industrial robots with the gripper having soft and sensored fingers was considered. Instead of using the well-known methods of reactive control based on force sensors at the manipulator wrist, a new method based on learning strategy by neural networks and pressure sensors on the soft fingers was proposed. The problem of learning reactive control was exemplified by the "peg-in-hole" task. Experiments were carried out using the industrial robot ASEA Irb-L6 with the Belgrade-USC-IIS Robot hand Model II whose fingertips are equipped with pressure sensors.

As a second example, the special genetic-connectionist algorithm for compliant robotic tasks is presented [163]. The systematic connectionist controller design approach is proposed to guarantee stability and desired performance of the robotic system for compliant tasks by effectively combining genetic algorithms (GA) with neural classification and neural learning control techniques. The effectiveness of the approach is shown by using simple and efficient decimal and binary GA optimization procedures to tune and optimize the performance of the neural classifier and controller, together with tuning of the feedback controller. The feedback gains of feedback controller, the weighting factors of the neural controller and the network topology of the neural classifier are tuned by the GA optimization procedure to achieve good results for force tracking of manipulation robots in contact tasks based on suitable fitness functions. In order to demonstrate the effectiveness of the proposed GA approach, some compliant motion simulation experiments with the robotic arm placed in contact with a dynamic environment have been performed.

As a third example, the connectionist approach to robotized vehicles (automobile) is presented. The aim of the proposed connectionist controller is to guarantee the desired performance of an advanced vehicle system. A supplementary neuro-compensator is proposed to ensure control system robustness and better controller adaptability to system uncertainties and model innaccuracies. This neural compensator is a part of the integrated active control algorithm based on the centralized dynamic control strategy and full vehicle model. Fast convergence of the learning process is achieved using the standard back propagation method. By introducing the supplementary neuro-compensator, greater robustness of the control system of the vehicle was achieved. The validity and effectiveness of the proposed method, based on adaptive capability of the neural compensator for a four-wheel steering system, have been demonstrated by simulation experiments.

2.   Connectionist Reactive Control for Robotic Assembly Tasks by Soft Sensored Grippers

In spite of the fact that the problem of assembly is one of the most important and most investigated working operations in the industrial application, it has not yet been generally resolved. One of the reasons is that the tolerances of the parts to be assembled are in many cases smaller than the accuracy of the contemporary robots. Because the robotic grippers in the manufacturing industry grip the object in a stiff manner, no relative motion of the object in the gripper is allowed. The tolerances of the assembled parts are the same tolerances that limit the motion of the gripper to achieve a desired position and orientation. Small deviations in the object orientation and position can cause significant reaction forces, especially if narrow tolerances are involved. Various approaches have been employed to overcome this problem. One of the approaches which uses the benefits of passive compliance was proposed in [35]. The gripper fingers are equipped with pressure sensors and covered with soft material, allowing object motion while it is being grasped. Changes of the pressure on the contact surfaces are sensed and used as basic information for corrective action during the task realization.

The basic principles of the peg-in-hole task realization involving a cylindrical peg are very simple, and the direction of corrective action can be successfully derived from the information obtained from the pressure sensors. Neural networks are very convenient for such tasks, and their use in the realization of the peg-in-hole assembly task is proposed.

Practical working conditions of the peg-in-hole realization involve various sources of uncertainty and noise, so that they can substantially degrade the performance of traditional (off-line planing) control methods. Hence, reactive control methods try to cancel the effects of uncertainty and noise by using autonomous on-line learning procedures based on the repetition of the working task. The control goal of learning is an on-line modification of the control signal yielding a whole set of corrective movements to achieve a successful task realization. In the process of reactive control, compliant behaviour of the system is accomplished either actively (using contact forces or tactile stimuli) or passively, by the inherent physical characteristics of the robot.

Hence, an approach to learning reactive control strategy for peg-in-hole insertion task using a manipulation robot with soft sensored fingers is proposed. The use of soft fingers in robotized assembly tasks enables self-adaptation of the manipulated object during the assembly operation. The main idea of the proposed method is the use of an array of separate pressure sensors on the surface of the robot's fingers, which give a reduced set of information needed for the choice of direction of corrective movements in the assembly task. The new method of corrective actions is based on representation and learning of nonlinear compliance laws. In this case, the nonlinear compliance is treated as a nonlinear mapping from the pressure sensors to the corrected motion. This nonlinear mapping is accomplished by the learning process based on the application of multi-layer neural networks.

2.1 Analysis of the assembly process with soft fingers

In contrast to the present practice, where the gripper's surface in contact with the object is hard and allows no relative motion of the object in the gripper, it is proposed to use a soft material between the hard object and the gripper's "skeleton". Thus the object is not gripped in the way that the system gripper-object can be considered as a single rigid body, but can move relative to the gripper, while the gripper is fixed in space when an external force is applied. Such a gripper with "soft" fingers can be used as a device with the characteristics of passive compliance. When the robot is used in tasks where the end effector is in contact with the environment, fast adaptation (compliance) to the external disturbance forces is indispensable. A change in the position of the object within the gripper causes a change of the finger elastic material stress state. Instead of measuring the reaction force at the robot wrist, the information is obtained from the stress state on the contact surface between the object and the gripper elastic pad. The force acting on each soft element is resolved in three mutually orthogonal directions. Two of them are parallel to the soft pad plane, while the third one is orthogonal to it. Thus, the force components are the normal component acting in the orthogonal direction with respect to the contact surface and two tangential components. One of the tangential components is selected to act along the gripper longitudinal axis, while the other is orthogonal to it. The normal component produces pressure between the object and the gripper, while the first tangential force in most cases corresponds to the insertion force, i.e., the force which has to be applied by the gripper to effect execution of the peg-in-hole assembly task. The other component of tangential force is produced by object rotation around the gripper roll axis.

The influence of the normal force component represented by the pressure between the object and the gripper is sensed by a set of pressure sensors . They do not provide information about the exact intensity and direction of the force applied but on the pressure profile shape and its change, which have a very rich information content. The two other tangential stress components have the character of force and torque, and they may be sensed in some other way, if necessary.

The basic information is the pressure portrait on the gripper fingers. The pressure portrait is established immediately when the object is gripped, even before the task execution starts. Let this portrait be called the initial pressure portrait. It is easy to see that any contact involving the free object will cause a change in this portrait. If the contact occurs as a consequence of a desired action, the pressure portrait obtained will be nominal. A good example is the writing operation. When the pencil is gripped, the pressure portrait corresponds to the initial one. When contact with the paper is established, the pressure portrait is transformed into the nominal one. In some cases, when the contact forces are unwanted (as in the assembly task) the nominal pressure portrait coincides with the initial one. Thus, a correct understanding of the pressure information is the key to a successful task realization.

The whole control strategy is based on the information about pressure profiles and their dynamic behaviour, knowledge of task realization and previous experience. For a known strategy and for a particular task realization, a direct relationship between the sensed stress state and the compensating action performed has to be established.

2.2   Assembly process

The most important problem in assembly processes is the forces which may arise at the contact points. Let the "peg-in-hole" task be adopted as a general representation of assembly tasks, and suppose that the object has already been gripped and brought close to the desired (nominal) position. Practically, this means that the object has been brought somewhere into the chamfering cone. Then in the initial movement of joining, two types of deviation may simultaneously occur:

a) Radial displacement of the object (peg) tip with respect to its nominal (desired) position at the beginning of insertion.

b) The peg angular deviation with respect to the hole axis.

Let us further define the nominal object position in the initial moment as the position which enables the realization of the assembly task only by moving the object in the direction of its axis, without any additional correction. This situation occurs when the axes of the peg and the hole coincide. If nominal conditions are preserved during the insertion, the assembly may be realized without any resistant force. Because it is hard to realize such perfect positioning, the object and the hole come in contact. A compensating (corrective) action of the robot has to be performed in such a manneres to eliminate or minimize deviations and enable successful completion of the assembly task.

The first contact between the object and the hole may occur in the cone formed by chamfering. Due to the very high stiffness of the arm of the industrial robot, small positional inaccuracies can produce very large reaction and large friction forces. It is not reasonable to suppose that the object tip will "naturally" slip along the surface of the chamfering cone. The main requirement imposed on compensating action should be to prevent the occurrence of large friction forces. This will be the case if no high insertion force is applied. Thus, the proposed action, which has to enable progression of the object into the hole, is the application of very high pressure in the direction of insertion, combined with the appropriate compensation action. The rotation of the peg around the contact point toward the cone surface until the reaction force comes out of the friction cone must be realized and sliding along the surface of the chamfering cone continued. The phase is over when the object tip enters the hole. To continue insertion, a different compensation strategy has to be applied. Due to imperfections in the relative positioning, narrow tolerances, and especially due to the gripper's soft contact surfaces, two-point contact between the peg and the hole will always occur. In this phase, when the object tip progresses into the hole and the two-point contact is established, the compensating movement is aimed at reducing or eliminating the friction force at the contact points. A deblocking movement has to be performed in the direction of motion generated by the reaction forces. The movement is limited by the size of the possible peg inclination with respect to the hole axis. When the object is inserted to its full depth, the motion of the object's gripped end is reduced to the level of manufacturing tolerances. Thus, the gripper compensation motion is not a function of time, but of the depth of the realized insertion, and this is extremely inconvenient for on-line measurement and for use as feedback information.

2.3 Learning compliance methodology by neural networks

The robotized assembly task can be considered as a combination of programmed motion and corrective actions in accordance with the information from the array of pressure sensors on the robot fingers. In this approach, corrective actions are based on representation and learning of the nonlinear compliance strategy. This nonlinear compliance is treated as a nonlinear mapping from the pressure sensors to a corrected motion. The nonlinear mapping is accomplished by the learning process, based on the application of a multilayer perceptron. The connectionist approach does not require explicit representation of compensation strategies, but the strategies are learned through the iterative presentations of input-output pairs. The type of desired compensation motion is defined in advance, on the basis of studying the process to be realized or based on experience. The primary aim of the compensating motion is to enable slip of the object tip along the cone surface toward the end of the hole. The newly-proposed method allows a fast and smooth control action without unwanted intermittent actions. A real peg-in-hole task also involves great sources of uncertainty and noise, which can be efficiently compensated for through the neural network model.

Synthesis of efficient learning reactive control is a multi-stage process, where a knowledge acquisition process is necessary as a preliminary phase. In this phase, the assembly task is realized by the operator's commands, based on visual information. In teaching compliance, the operator demonstrates how the motion should be corrected in each training situation. During this phase, the intensive acquisition process of input network data (pressure information) and output network data (corrective actions) is realized. The training samples can be acquired using some other methods, such as measurement of human motions, recording of operations of a master-slave manipulator, or by simulation and off-line programming.

After that, in the process of off-line learning using the previously collected input-output pairs, the three-layer perceptron must find a generic form of association of pressure with the a priori defined corrective actions. In order to enhance speed of the learning process, special rules, based on the recursive least squares approach, are used [168]. In the final stage, the robot works in an automatic mode, without interventions of the human operator and with the corrective strategy, which the neural network gives as a result of the on-line pressure sensor information. This corrective strategy is defined by the generic knowledge incorporated through the fixed values of weighting factors of the proposed three-layer perceptron. The neural network must be capable of high-level generalization for input-output pairs which are not captured in the process of operator training. Also, by this on-line process, updating and improving the nonlinear mapping represented by the nonlinear network could likely be performed by the robot hand.

2.4   Experimental results

The connectionist approach was experimentally verified on the example of an assembly task of the peg-in-hole type. Two basic actions had to be realized simultaneously: a small force acting constantly in the direction of the object axis, and adequate compensating movements performed according to the actual pressure profile.

A Belgrade-USC-IIS multifingered hand Model II (Figure 8.1) with three sensored fingers was used as gripper. The thumb and finger were equipped with a single tactile sensor covering the whole fingertip, while the tip of the middle finger was sensored with six separate sensors placed as strips orthogonal to the finger axis. Such a design of pressure sensors enables the pressure profile on the fingertip to be sensed and recorded. It is clear that it would be more desirable to sensor all fingertips in the same way as the middle finger, but this was not possible due to the limited number of sensor inputs available. As the sensor's presence should not affect the finger's local shape adaptability. use was made of FSR tactile sensors. which consist of two thin foils. On one of them, conductive pattern is printed, while the second one is covered with a thin film of FSR material, which changes its resistance in response to the applied force. It is clear that such a sensor enables only the normal component of the contact force to be measured, while the force in the direction of the object axis cannot be controlled in this way.

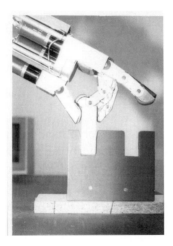

Figure 8.1. Belgrade-USC-IIS multifingered hand Model II

To ensure constant action of the force in the direction of insertion, a specially designed device, schematically shown in Figure 8.2, was realized. Instead of utilizing the gripper with the grasped peg to ensure pressure in the insertion direction, this task was assigned to the object with the hole in it. Hence, this object was fixed to the end of a metal rod, which can slide along the fixture in a vertical direction. The other end of the metal rod has connected to the rubber strips. When an external vertical force is applied onto the object with the hole, it moves downward until the force is balanced by elastic forces generated by the rubber strips. When the peg is in the contact with the hole, and the robot is performing a compensating action to eliminate relative displacement in the peg position (while the peg axis is in the cone), the rod will move upward, and partial insertion will be automatically performed. In this way, full depth insertion may be realized. The robot hand was fixed to an industrial robot ASEA Irb-L6 with

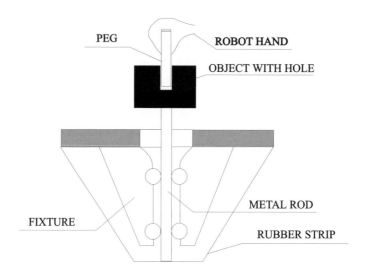

Figure 8.2.   Experimental fixture

standard industrial controller. The robot hand controller was a PC computer with two additional boards. Communication between these two controllers was realized via an ASEA standard Computer Link Board.

Because direct access to the robot controller was not possible, but only via the Computer Link Board, a special technique was developed to accomplish compensation movement in the desired direction. Any inclination of the peg produces a unique pressure pattern on the fingertips, which requires hand movement in the proper direction. The object with hole was placed in the centre of the circle (Figure 8.3). In the general case, an arbitrary direction of the peg relative to the hole axis will cause a unique change of the pressure on the contact surfaces of the fingers. Because of the specific hand design, the fingers are not able to grip the peg at equidistant points of its cross-section. As for the hand, two fingers are parallel, with the thumb being on the opposite side of the peg. Thus only some limited angular deviation relative to the direction defined by points 2 and 5 in Figure 8.3 may be successfully handled. Then, instead of covering the whole circle, only six target points (1-6) are defined in advance. In the teaching phase, the operator, with the aid of joystick, brings the peg and the hole into contact. Instead of calculating the angle of direction in which the

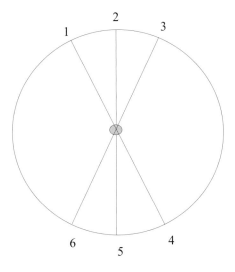

Figure 8.3. Selected compensation directions

hand is to move, the operator on the basis of visual inspection, defines one of the points (1-6) and the robot moves in this direction. During this motion, partial (or full) insertion is realized. When jamming occurs, the motion is stopped. The operator now eliminates the direction of hand motion that caused jamming, and the procedure is repeated. Each time, all the pressures from the sensors are recorded, as well as the corresponding action chosen by the operator. This set of data is used to train the neural network.

The overall structure of the multi-layer perceptron for learning the pressure-direction relations is presented in Figure 8.4. The input layer of the network consists of normalized signals from the pressure sensors (eight pressure sensors on the three soft robot fingers plus the bias member). The hidden layer has 32 neurons plus the bias member. The network output produces 6 different control signals for the a priori defined compensation actions. The logistic sigmoid function was used as activation function. The process of learning has accomplished through iterative presentations of the training patterns (71 samples of input-output pairs) acquired in the preliminary phase. During learning, connection weights (matrices W1 and W2) are adjusted until the appropriate level of output error is achieved. The process begins with random connection weights. For the process of learning, rules based on the recursive least squares learning procedure [168] are chosen because of their fast convergence properties. The results of learning (square error criterion) are presented in Figure 8.5. After learning, the values of the connection weights are stored in the robot controller memory for use in the on-line control. During process of on-line control, the robot

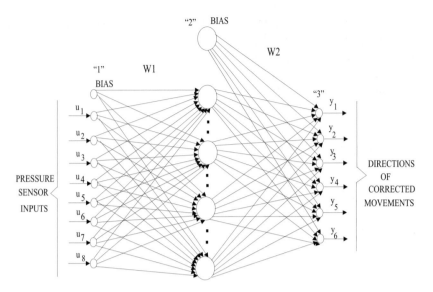

Figure 8.4.   Structure of multilayer perception used in off-line learning procedure

controller uses the stored connection weights, takes a forward pass through the neural network and makes decisions about compensatory action based on the value of the network output. The results obtained during the on-line control

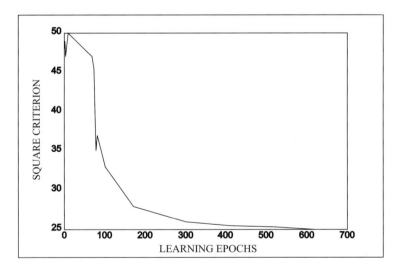

Figure 8.5.   Square error criterion during process of off-line learning

show that the neural network is capable of accurately performing the working task, using high-level generalization for the input-output pairs, not captured in the process of the operator training.

Finally, a gripper with soft fingers with the characteristics of a passive compliance is used for the realization of the peg-in-hole assembly task. The contact force acting on the object's free end causes object displacement within the gripper, resulting in stress changes at the elastic contact surface. The pressure profile is sensed by a set of pressure sensors placed under the soft finger tissue. A compliance control strategy for the assembly task, using robot hands with soft fingers and the fast learning capabilities of connectionist structures, are developed. The new method of corrective actions is based on representation and the multi-stage learning of nonlinear compliance between the information from the pressure sensors and the corrective motion, using the multi-layer neural network. The major contribution is a new learning compliance strategy, based on the relations between the pressure sensors on the robot hands and the direction of the corrective actions.

3. Genetic-Connectionist Algorithm For Compliant Robotic Tasks

In this example, the idea was to enhance the capabilities of the proposed connectionist control algorithm for contact robot tasks [170] [171] by synthesizing new control algorithms based on the genetic tuning of nonlearning control part (conventional force PI controller) and learning control part (neural "off-line" classifier and neural "on-line" controller). In the proposed nonlearning control part we use control algorithms for stabilizing the interaction force based on the quality of the transient processes [356]. These stabilizing control laws ensure exponential stability of the closed-loop system. However, the property of closed-loop stability is a very basic requirement, but it is never the only one. In addition to stability, the user is usually very interested in adapting the performance of the system in terms of overshoot, oscillation and settling time. For many such controllers there are currently no systematic approaches to choose the control parameters yielding the desired performance. The controller parameters are usually determined by trial-and-error through simulations and long-lasting experimental tests. In such cases, the paradigm of GAs appears to offer an effective way of automatic and efficient searching for a set of controller parameters yielding better performance. There are several efficient GA methods proposed, intended for various special purposes in robotics [123][92] [3][49] [86].

Compared with conventional optimization methods, GAs possess many advantages (global, data-independent and robust method). Furthermore, GA can be directly applied to solve an optimization problem with a certain fitness function without reformulating the problem into a suitable form. Here, a simple GA variant which works directly on real (decimal) parameters, is used. Decimal-type GAs are equivalent to the traditionally used binary-type GAs in optimization. The real-type GAs for computer-based numerical simulations lead to high computational efficiency, smaller computer memory requirements with no reduction of precision and to greater freedom in selecting genetic operators.

A systematic approach to the design of controllers for both closed-loop stability and desired performance by using GA to tune the position and force feedback gains is proposed. GA utilized is selected to be of the decimal real number type, to achieve simple and efficient computation. Two types of fitness functions are considered for optimization of the controller performance: integral of squared errors (ISE) and integral time-multiplied absolute value of errors (ITAE).

In order to improve the convergence process (a time-consuming process of learning because of a nonsystematic approach to the determination of the network structure), an efficient GA is proposed to choose the appropriate topology of the multi-layer perceptron that performs neural classification. Also, in order to improve the learning process, GA optimization is used to determine the weighting factors for neural compensation of the robot dynamic model in the on-line control algorithm. In both previous cases, binary representation of the problem is used to achieve good results for the robot compliance task, based on suitable fitness functions.

A control algorithm, based on stabilization of the interaction force with a preset quality of transient responses, is considered, which has the following form [356]:

$$\tau = H(q)M^{-1}(q)[-L(q,\dot{q}) + S^T(q)F] + h(q,\dot{q}) +$$
$$J^T(q)\{F_p - \int_{t_0}^{t} [KFP\mu(\omega) + KFI\int_{t_0}^{t} \mu(\omega)dt]d\omega\}$$

$$(8.1)$$

where $\mu(t) = F(t) - F_p(t)$; $KFP\epsilon R^{n \times n} = diag[KFP_{ii}]$ is the matrix of proportional force feedback gains; $KFI\epsilon R^{n \times n} = diag[KFI_{ii}]$ is the matrix of integral force feedback gains. Here, it has been assumed that the interaction force in a transient process should behave according to the following differential equation:

$$\dot{\mu}(t) = Q(\mu) \qquad (8.2)$$

$$Q(\mu) = -KFP\mu - KFI \int_{t_0}^{t} \mu dt \qquad (8.3)$$

The PI- force regulator (continuous vector function Q) is chosen so that the system defined by (8.2) will be asymptotically stable as a whole.

3.1   GA tuning of PI force feedback gains

In order to further simplify the genetic process, the set of tuning force gains KFP and KFI is reduced to the single parameter ω_n, where ω_n is the natural frequency of the second-order linear system defined by the characteristic equation:

$$\mu_i(\omega) + \int_{\omega_0}^{\omega} [KFP_{ii}\mu_i(\omega) + KFI_{ii} \int_{t_0}^{t} \mu_i(\omega)dt]d\omega = 0 \qquad (8.4)$$

Previous forms of the characteristic equations are equivalent to the following equation:

$$\ddot{\eta}_i + 2\zeta\omega_n\dot{\eta}_i + \omega_n^2 = 0 \qquad (8.5)$$

If we assume for the second-order system that the critical damping $\zeta = 1$, the feedback gains are given by

$$KFP_{ii} = 2\omega_n \qquad (8.6)$$

$$KFI_{ii} = \omega_n^2 \qquad (8.7)$$

In this way, only natural frequency is chosen for genetic tuning. The initial population of the size N is generated randomly to start the optimization process. The total population of each generation is evaluated using a suitably chosen performance criterion (ISE or ITAE). Reproduction as the primary genetic operator is based on using the best $N/2$ individuals of the current generation to be the parents of the next generation. A weighted-average cross-over genetic operator based on decimal numbers is applied [92]. From the parents ω_{n1} and ω_{n2}, two new descendants are generated by the following terms:

$$\omega_n^1 = r * \omega_{n1} + (1 - r) * \omega_{n2} \qquad (8.8)$$

$$\omega_n^2 = (1 - r) * \omega_{n1} + r * \omega_{n2} \qquad (8.9)$$

where $r\epsilon(0, 1)$ is a random number. Mutations are based on the following changes of natural frequency:

$$\omega_n^1 = \omega_n + (r - 0.5) * 2 * \Delta\omega_n^{max} \qquad (8.10)$$

where $\Delta\omega_n^{max}$ is the maximum change of natural frequency.

The objective of the GA optimization is to obtain better end-effector performance, i.e., to find the PI force feedback gains as fast as possible with minimal oscillation and overshoot. The fitness functions are defined according to the following equations:

$$ISE = \int_0^T \mu^2(t)dt \tag{8.11}$$

$$ITAE = \int_0^T \mid t\mu^2(t) \mid dt \tag{8.12}$$

3.2   Simulation experiments involving genetic algorithms

To demonstrate the performance of contact control schemes with GA tuning of the general controller and neural elements, compliance control implementations are simulated using the robot MANUTEC r3 in contact with various models of robot environment. Complete parameters of the robot and its environments are given in [171].

The technological working demands for the reference working operation are defined as follows: a) The working tool of the robot is realized in the form of a rotational-milling tool, performing surface processing a plane which is parallel to the $X - Y$ plane; b) The tool trajectory is 100 $[mm]$ long; c) The task of the robot is to carry out the machining process of the work surface along the prescribed trajectory with a desired contact force $F_N^0 = 5\ [N]$ and a prescribed velocity of 25 $[mm/s]$. The following initial conditions were used in the simulation: the robot gripper starts with zero initial velocity; the settling time of the desired contact force is given as $t = 0.5\ [sec]$.

To investigate the effect of the GA optimization procedure for tuning the PI local force gains, simulation experiments were conducted with the appropriate initial set of PI local force gains. It is necessary to specify the range of the controller parameter (natural frequency). Including the maximal torque value given by the actuator limits, we obtain the range of the natural frequency:

$$0 < \omega_n \le 32. \tag{8.13}$$

In the simulation, the population size of each generation is set to be N=40. The maximum mutation values for $\Delta\omega_n^{max} = 0.5$. The evaluation process is terminated when the change of fitness function is small in a certain number of successive generations. The results of the GA optimization procedure are shown in Figures 8.6 and 8.7.

It is obvious that better performance (corresponding to smaller values of the fitness functions) will be obtained with the progression of the GA process.

Based on the previous GA optimization, PI force gains are synthesized using the same system frequencies for all different working environments ($\omega_n = 11.86Hz.$). In the phase of connectionist off-line training, an efficient GA is used in order to select the optimal topology of the neural network (described

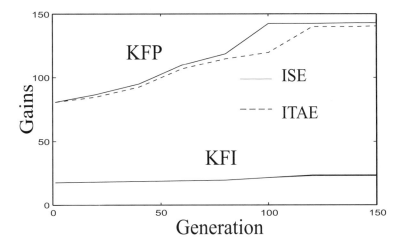

Figure 8.6. Best force feedback gains KFP and KFI according to ISE and ITAE criterions

in subsection 4.1.1). Using the adopted GA network topology and the learning process, the training is carried out with the stored weighting factors. In the generalization test, the "off-line learned" and GA tuned neural classifier is included into the control algorithm for recognition of the unknown robot environment. In a similar way, the GA approach, based on binary representation is applied to determine the weights of the second neural network for compensating the robot uncertainties.

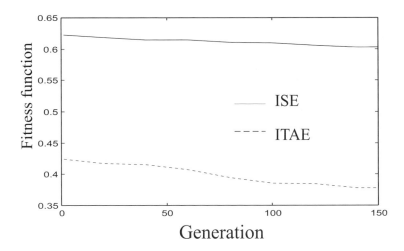

Figure 8.7. Best values of ISE and ITAE criterion during evaluation process

4.   Connectionist Compensator For Advanced Integrated Vehicle Controller

The contemporary automotive industry adapts to the traffic demands by endeavouring to improve the vehicle's dynamic performance, ride quality, and ride safety. The evolution of road vehicles as transport means is directed towards improving their manoeuvring capabilities and increasing their automation level. Thus, the highest level of automation with road vehicles is the synthesis of their integrated active control systems. Similar control systems have existed for a long time in airplanes, submarines, different kinds of robots and other large-scale dynamic systems. Full automatic control of motion and road vehicle performance is still in an experimental phase. Some satisfactory results have been achieved in the control of vehicles operating with automated highway systems, i.e., with intelligent transportation systems [30]-[273]. The hybrid neuro-dynamic controller of road vehicle described here differs essentially from other solutions known in the open literature. Such a controller simultaneously takes into account the complete system dynamics in the all main directions of vehicle body motion: longitudinal, lateral, vertical, as well as the directions ofthe roll, pitch and yaw angles. The centralized approach to control, by using the entire spatial model of the vehicle, makes the proposed controller advantageous over similar control schemes applied in the motion control of autonomous vehicles with strongly expressed dynamics. Such a case appears during a ride along the path at a rather high velocity, sudden variations of the vehicle course during motion, or with significant variations of the road geometry parameters.

The road vehicle, as an object of automatic control, represents a complex dynamic system consisting of interconnected rigid and elastic bodies. A vehicle, as a real rigid-body system, possesses over 30 DOFs of motion. From the standpoint of the conventional spatial vehicle model [225], two functional modules can be distinguished: the *vehicle body (structure)* and its *suspension system*. Thus the vehicle body will be considered as a mechanical system representing a rigid body supported by its elastic suspensions [273]. The vehicle suspension system and the ground over which the vehicle moves, representing its "dynamic environment", will also be considered. The choice of the appropriate control strategy and the manner of its realization represent a delicate problem, solution of which demands sufficiently deep knowledge of the dynamic behaviour of the road vehicle under various motion conditions. Thus, a sufficiently faithful vehicle model is needed. The models used most frequently are simplified system models such as, quarter-car model [138], half-car model (so called "bicycle model") [275], planar model of lateral dynamics [137], model of longitudinal dynamics [305, 116], etc. Here, we present a relatively complex, nonlinear, spatial model of road vehicle, which enables a centralized approach to dynamic control. The adopted model is a good approximation of the considered real system, so that it describes sufficiently well the nonlinear nature of the vehicle dynamics in a rather broad domain of state variables variation and control/disturbance signals.

Previous solutions for vehicle controllers have assumed the synthesis of independent partial controllers in particular directions of vehicle motion and their simple summation into the so-called *summated controller*, designed without testing the system's stability. Knowing that in some cases (as, for example, during motion at a higher forward speed), when a strong coupling of dynamic effects within the system between particular directions of vehicle motion takes place, it may happen that the particular control actions come into collision. Instead of a simple superposition of control signals in the summated controller and with the aim of avoiding the negative mutual influence of control actions in particular directions, we propose the synthesis of the so-called *integrated vehicle controller*, based on the centralized dynamic model of the system. By appropriate choice of control parameters of the integrated vehicle controller it is possible to ensure the desired stability of the considered system in the presence of external perturbations and with the parameter variation in a predicted range.

However, the mathematical model of the road vehicle used in the synthesis of the vehicle controller describes only the most significant dynamic effects that exist in the system during motion. As a consequence, some dynamic effects such as elastic modes in the vehicle mechanism, dynamics of the vehicle actuators, time delays in the drive units and control subsystems, real friction at the joints of the suspension system, fluid stiction in the hydro-cylinders of the viscous absorbers, etc., were not explicitly modelled. These phenomena influence the overall dynamic behaviour of the road vehicle during motion. For the mentioned reasons, in order to control the vehicle while ensuring full controllability and stability on the road, it was necessary to add a supplementary neuro-compensator to the existing control structure of the dynamic controller to compensate for the corresponding system uncertainties.

Recently, neural networks have aroused considerable interest in the areas of identification and control of advance vehicle systems [269], [182], [254], [257]. In general, neural networks can be applied for vehicle control as a neural network-based fully controlled system and as a partially supporting tool for the existing control system. The essential idea of the proposed active integrated vehicle control is to reduce to the design of a centralized hybrid (neuro-dynamic) controller, which is robust to the parametric and structural inaccuracies of modelling, as well as the action of stochastic external disturbances of foreseeable intensity. The omnipresent structural and parametric model inaccuracies will be compensated a posteriori by designing a supplementary neural compensator, which takes into account the existing uncertainties in the system.

4.1   Model of the vehicle dynamics

The complex nonlinear model that Peng and Tomizuka applied in their simulation experiments [273] was used in the synthesis of the road vehicle dynamic
controller. This model was rearranged and extended taking into account the
dynamics of the suspensions and tyres. A nonlinear vehicle model having 22
DOFs, with the possibility of autonomous 4-wheel driving (4WD) and 4-wheel
steering (4WS), was used for the stability analysis, control synthesis, parameters estimation, and simulation [283]. The model describes the motion of the
vehicle mass centre (MC) in 3 coordinate directions (x, y, z) and 3 rotations $(\phi,$
$\theta, \varepsilon)$ of the vehicle about its main axes of inertia. The model also describes the
dynamics of the 4-wheel suspension system in the vertical direction, as well as
the tyre dynamics in the same direction. Each wheel, in addition to the vertical
tyre deflection, possesses 2 extra DOFs: rotation about the horizontal axis at an
angular velocity ω_i (for $i = 1, 4$) and, rotation about the vertical axis δ_i w.r.t.
the road surface. The second rotation represents a change of the tyre's ground
steering angle.

Instead of a 4-wheel suspension of the vehicle body, we assume that the
vehicle structure can rely upon only one imaginary suspension system, which
is in contact with the vehicle body at one suspension system joint. In that sense,
the imaginary suspension system shown consists of one linear and two torsional
spring dampers and the corresponding viscous dampers. The equivalent model
resembles a typist's chair, so we call it the "typist's chair" model of the road
vehicle. Assuming the vehicle model to be in the "typist's chair" form, the
vehicle structure can be treated as a rigid body, which is supported only at one
point in its "dynamic environment". In this way, all external resistances, as well
as the reaction forces/moments acting upon the vehicle body, can be relatively
easily reduced to the mass centre (MC).

The vehicle body model is determined by its rigid-body dynamics, and it can
be expressed via the vector equation [283]:

$$H(q, d)\ddot{q} + h(q, \dot{q}, d) = \tau + F(q, \dot{q}, d) \tag{8.14}$$

where:

$$q = [x \ \ y \ \ \varepsilon \ \ z \ \ \phi \ \ \theta]^T, \quad \dot{q} = [\dot{x} \ \ \dot{y} \ \ \dot{\varepsilon} \ \ \dot{z} \ \ \dot{\phi} \ \ \dot{\theta}]^T$$

are (6×1) vectors of the system state variables describing the position/orientation
and velocity of the vehicle body MC with respect to the coordinate system fixed
to the ground; x, y, z are the longitudinal, lateral and vertical positions of the
vehicle MC along the three coordinate directions, expressed in $[m]$; $\phi, \theta, \varepsilon$ are
the corresponding angles of roll, pitch and yaw of the vehicle body in [rad];
$H(q, d)$ is a (6×6) inertia matrix, expressed in $[kg]$ and $[kgm^2]$, respectively;
$h(q, \dot{q}, d)$ is a (6×1) vector of gravitational and centrifugal forces acting at
the vehicle MC, expressed in $[N]$ and $[Nm]$, respectively; τ is a (6×1) vector

of driving forces and torques referred to the vehicle MC, expressed in $[N]$ and $[Nm]$, respectively; $F(q, \dot{q}, d)$ is a (6×1) vector of the external forces and torques acting on the vehicle body during its motion along the road. Elements of this vector take into account the forces and torques of tyre rolling resistance, aerodynamic resistance forces during motion, as well as the damping torque of the yaw rate during cornering. The vector d represents an $l \times 1$ vector of the system parameters.

The "dynamic environment" of the road vehicle is approximated by a model of the vehicle suspension system in a broader sense. Thus it is assumed that the dynamics of tyre-pneumatics in the "vertical" direction can be reduced to the equivalent vehicle suspension system dynamics. In the robotics literature, the dynamic environment is usually represented in the linear impedance form: $M\ddot{q}(t) + B\dot{q}(t) + Kq(t) = -S\,F$, taking into account the inertial (M), damping (B) and elastic (K) characteristics of the environment. In some contact tasks of the robot interacting with the dynamic environment, the environment dynamic model is adopted in the nonlinear form [60], [356]:

$$M(q, \tilde{d})\ddot{q} + L(q, \dot{q}, \tilde{d}) = -S\,F \qquad (8.15)$$

where $M(q, \tilde{d})$ is the (6×6) matrix describing the equivalent environment inertia in 6 coordinate directions, while $L(q, \dot{q}, \tilde{d})$ is the (6×1) nonlinear vector function, which takes into account the equivalent elastic and damping characteristics of the environment interacting with the vehicle body. Parameters of the "dynamic environment" are in general variable. They are defined by the vector $\tilde{d}$ of dimension ($\tilde{l} \times 1$). The transformation matrix S is a (6×6) matrix. It takes into account the relative orientation of the vector of the external forces and moments F, which act upon the environment w.r.t. the fixed coordinate system attached to the ground surface.

In accordance eith the decoupling of its dynamics in particular directions the dynamic environment model can be described by the following relation:

$$\begin{bmatrix} M_{11} & M_{12} \\ M_{21} & M_{22} \end{bmatrix} \begin{bmatrix} \ddot{q}^{(1)} \\ \ddot{q}^{(2)} \end{bmatrix} + \begin{bmatrix} L^{(1)} \\ L^{(2)} \end{bmatrix} = -S \begin{bmatrix} F^{(1)} \\ F^{(2)} \end{bmatrix} \qquad (8.16)$$

In order to simplify the modelling process it was assumed that the roll and pitch axes of the vehicle body pass through the vehicle body MC. In that case, the matrix S is identical to a sixth-order square unit matrix $S \equiv I$.

4.2 Control strategy

The purpose of integrated vehicle control is to ensure global motion stability of the road vehicle. The vehicle autopilot also has to ensure the satisfactory dynamic behaviour of the system, i.e., its satisfactory responses. Sudden changes of the magnitudes of longitudinal and lateral acceleration, i.e., personal feeling of jerky motion, must be avoided. For the ride comfort, it is very important that the synthesized controller ensures sufficient suppression of the vehicle body vibration due to bobbing motions, as well as the desired minimization of the deflections of the suspension system and tyre pneumatics. The control strategy proposed here is the strategy of the so-called *distributed hierarchy control*. An example of its implementation was described in [266]. The solution presented here is based on knowledge of the entire vehicle dynamics. Thus, this control strategy can be named as the *strategy of centralized dynamic control* . The term "hierarchy" is used because the object control is realized on two levels: *tactical* and *executive*. They are separated w.r.t. the existing dynamic modules: the *vehicle body* and *vehicle active suspension*. The control is "distributed" because the global control signals that are generated on the higher level are distributed as reference signals on the lower, executive control level.

On the higher, tactical control level, the vehicle controller, which is based on the information about errors of global state variables q and $\dot{q}$, calculates the vector of control forces and torques which have to act at the vehicle MC to realize the desired motion. The control forces and torques calculated in this way are realized indirectly by the action of the vehicle end-effectors operating in the frame of the executive control level. For full automatic control of the vehicle motion and performance, various types of actuators can be applied. They operate with the scope of the particular vehicle active systems, such as active suspension system, active steering system, and active traction system . The mentioned active systems can be synchronously applied on all four wheels, which should give high control performance and enable full automatic control of the autonomous vehicle on the road.

The vehicle autopilot demands exact and timely information about positions of the vehicle on the road, its relative velocity, distance from static and mobile obstacles along the trajectory and changes of the road geometry parameters. Important prerequisites of the application of the autopilot system structure to be described in the text to follow are: (I) Nominal vehicle trajectory has to be known in advance or has to be generated in real time during motion; (II) At any time instant, it is possible to measure the current position of the vehicle body MC sufficiently exactly w.r.t. the road central line as a nominal path. It is also assumed that there are appropriate sensors on the vehicle that measure the relative attitude deflections and payload magnitudes at the vehicle MC; (III) It is possible to measure, or estimate in some way, the time-dependent magnitudes of the vehicle body velocities in all six coordinate directions of motion; (IV) Dynamic models expressed by the relation (8.14), by its structure and complexity, describe sufficiently well the main dynamic effects in the system; (V) Estimation of the vehicle dynamic parameters and parameters of interaction between the tyre pneumatics and road surface is realizable in real time.

4.3 Synthesis of the vehicle dynamic controller

The dynamic controller of the road vehicle is synthesized on the tactical control level. The magnitudes of the control forces and torques are calculated on this control level. They have to act at the MC of the vehicle body in order to realize a desired motion. These variables are realized by the actuators of the vehicle active systems, whose control signals are calculated on the executive level [283]. Relating to the nature of the deviation error signal (position/velocity or force/moment) in the feedback loop of the control scheme, two different types of dynamic control algorithms can be synthesized: a *pure position control law* and a *combined position/force control law*. Which of these two control laws will be implemented in the scope of the vehicle dynamic controller depends on the concrete control requirements. Pure position control is relatively simple for synthesis because it demands only information about the exact position and velocity of the vehicle body relative to the road surface. On the contrary, in the case of implementation of the combined control law, it is not necessary to measure the attitude deflections of the vehicle body directly. Instead, it is necessary to measure the corresponding forces and moments acting at the MC of the vehicle body w.r.t. its equilibrium state at rest.

On the basis of experience with road vehicles as large-scale dynamic systems it should be pointed out that the dynamic interconnections inside the vehicle's mechanism are not of equal intensity in some particular directions. Thus, the longitudinal, lateral, and yaw motion of the vehicle body are mutually strongly coupled. Moreover, the heave motion of the vehicle body during riding is directly dependent on the corresponding displacements in the roll and pitch directions. This is the reason why the considered system can be dynamically decoupled in two dynamic modules: (I) the vehicle dynamics *"in the plane of the road surface"* and, (II) the vehicle dynamics *"in the conditionally vertical plane"*. Dynamic interconnections between DOFs within the mentioned dynamic modules are strongly expressed, while the interactions between these modules are relatively weak. Bearing in mind the previous remarks and taking into account the fact that the system stabilization in longitudinal, lateral and yaw directions (x, y and ε direction) is the most important task of the vehicle control, this naturally imposes the necessity of applying the position/velocity control in these motion directions. In the remaining three directions (vertical, roll, and pitch) it is appropriate to apply force and moment control. In this way, a uniform tyreload distribution upon the wheels is ensured, with an indirect positive influence on the entire system stability.

>From the standpoint of previous considerations, the nominal position vector q_0 is partitioned into two subvectors $q_0^{(1)}$ and $q_0^{(2)}$. The partition of the external forces vector F_0 is carried out in the same way. As mentioned above, the vehicle body possesses $n = 6$ motion DOFs in q-directions, so that the vector F of

external forces and moments acting upon the vehicle structure is also of the order $m = 6$. In n_1 directions ($n_1 < n$), the vehicle nominal trajectory $q_0^{(1)}$ is prescribed directly. In these directions, position and velocity are directly controlled. Simultaneously, in m_2 directions ($m_2 < m$), the variation function of the programmed force/moment $F_0^{(2)}$ is prescribed. In these directions the force/moment is controlled directly. The mentioned vectors $q_0^{(1)}$ and $F_0^{(2)}$ are of dimensions ($n_1 \times 1$) and ($m_2 \times 1$) respectively, and they are prescribed in advance as programmed (nominal) values. The remaining two subvectors $q_0^{(2)}$ and $F_0^{(1)}$ are of dimensions ($n_2 \times 1$) and ($m_1 \times 1$), and they are calculated indirectly by using the model (4). Bearing all this in mind, the vectors q_0 and F_0 can be defined in the following partioned forms:

$$q_0 = [q_0^{(1)T} \ q_0^{(2)T}]^T, \quad \text{where} \quad q_0^{(1)} = [x_0 \ y_0 \ \varepsilon_0]^T \ \text{and} \ q_0^{(2)} = [z_0 \ \Phi_0 \ \theta_0 \ ]^T \tag{8.17}$$

$$F_0 = [F_0^{(1)T} \ F_0^{(2)T}]^T, \quad F_0^{(1)} = [F_X^0 \ F_Y^0 \ M_Z^0]^T \ \ F_0^{(2)} = [F_Z^0 \ M_X^0 \ M_Y^0]^T \tag{8.18}$$

Elements of these vectors belong to the set of real numbers: $q_0^{(1)} \in R^{n_1 \times 1}$, $q_0^{(2)} \in R^{n_2 \times 1}$, $F_0^{(1)} \in R^{m_1 \times 1}$, $F_0^{(2)} \in R^{m_2 \times 1}$. For the considered object of control, their dimensions are: $n_1 + n_2 = n$, $m_1 + m_2 = m$, $n = m = 6$, $n_1 = n_2 = 3$ and $m_1 = m_2 = 3$.

Dynamic Position Control Law:. If in the relation (8.14) q denotes the (6×1) vector of global state coordinates (position and orientation of the vehicle body), $q_0 = [x_0 \ y_0 \ \varepsilon_0 \ z_0 \ \phi_0 \ \theta_0]^T$ is the vector which denotes the desired, i.e., programmed, trajectory of the road vehicle. Hence, the position control law can be now expressed as:

$$\tau = \hat{H}(q, d) \, [\ddot{q}_0 + \Pi(\Delta q, \Delta \dot{q})] + \hat{h}(q, \dot{q}, d) - F(q, \dot{q}, d) \tag{8.19}$$

$$\Pi(\Delta q, \Delta \dot{q}) = -K_V \, \Delta \dot{q} - K_P \Delta q, \quad PD - regulator \tag{8.20}$$

$$\Pi(\Delta q, \Delta \dot{q}) = -K_V \, \Delta \dot{q} - K_P \Delta q - K_I \int_0^t \Delta q \, dt, \quad PID - regulator \tag{8.21}$$

where $\hat{H}$, $\hat{h}$ and F are the corresponding estimated or measured values of matrices and vectors of the model (8.14); K_V, K_P and K_I are 6×6 matrices of velocity, position and integral control gains respectively, in PD or PID variant of the chosen controller. The matrix of control gains K_P, K_V and K_I can be defined as in [283], for the case when the system parameters are ideally known. In other cases, i.e., when parameter uncertainty exists, the corresponding gains of the controller can be determined by applying the algorithm for the system practical stability test.

Dynamic Position/Force Control Law:. The dynamic position/force control law demands the calculation of position and velocity errors of the vehicle body MC w.r.t. their desired (nominal) values, as well as deviations of the external forces and moments acting at the MC. In the case of applying this control law it is not necessary to measure the state variables $q(t)$, $\dot{q}(t)$ in all coordinate directions but only in x, y and ε directions. Similarly, it is of importance with the components of force/moment vector $F(t)$, which have to be measured in the z, ϕ and θ directions. The basic idea of the combined position/force control law is the choice of position/velocity control in some chosen directions, while in the rest of the directions the force/moment control is applied. Practical benefits of the implementation of this control algorithm in the designed vehicle controller are shown by the fact that in some directions is easier to measure the force and moment than their positions and velocities. Separation of the directions is done under the criteria of *minimal dynamic coupling* between the dynamic modules. Thus, the vectors q and F from the model (8.14) can be formally partitioned in two subvectors

$$
\begin{aligned}
q &= [q^{(1)T}\ q^{(2)T}]^T, \quad F = [F^{(1)T}\ F^{(2)T}]^T, &\text{(8.22)}\\
q^{(1)} &= [x\ y\ \varepsilon]^T, \quad q^{(2)} = [z\ \phi\ \theta]^T, &\text{(8.23)}\\
F^{(1)} &= [F_X\ F_Y\ M_Z]^T, \quad F^{(2)} = [F_Z\ M_X\ M_Y]^T &\text{(8.24)}
\end{aligned}
$$

in the way it was done with their nominal forms in (8.17) and (8.18). Such partition of the vectors q and F is done because the influence of the forces/moments upon the corresponding displacements of the vehicle body ($F^{(1)}$ to the $q^{(2)}$ and $F^{(2)}$ to the $q^{(1)}$) are weak, and because the dynamic behaviour of the system in these directions can be mutually decoupled. In fact, this means that the position will be controlled in the directions x, y and ε (longitudinal, lateral and yaw) while in the directions z, ϕ and θ (vertical, roll and pitch) force/moment control will be applied.

Hence, the dynamic position/force control law can be defined in the following form:

$$\tau = \hat{H}(q,d)\ddot{q}_c + \hat{h}(q,\dot{q},\tilde{d}) - F \tag{8.25}$$

$$\ddot{q}_c = \begin{bmatrix} \ddot{q}_0^{(1)} + \Pi(\Delta q^{(1)}, \Delta\dot{q}^{(1)}) \\ -\hat{M}_{22}^{-1}[(F_0^{(2)} + \int_0^t Q(\Delta F^{(2)})dt) + \hat{M}_{21}(\ddot{q}_0^{(1)} + \Pi(\Delta q^{(1)}, \Delta\dot{q}^{(1)})) + \hat{L}^{(2)}] \end{bmatrix} \tag{8.26}$$

$$Q(\Delta F^{(2)}) = -K_F^{(2)}(F^{(2)} - F_0^{(2)}) - K_{FI}^{(2)} \int_0^t (F^{(2)} - F_0^{(2)})dt, \quad PI-regulator \tag{8.27}$$

$$Q(\Delta F^{(2)}) = -K_F^{(2)}(F^{(2)} - F_0^{(2)}), \quad P-regulator \tag{8.28}$$

$$\Pi(\Delta q^{(1)}, \Delta \dot{q}^{(1)}) = -K_V^{(1)}(\dot{q}^{(1)} - \dot{q}_0^{(1)}) - K_P^{(1)}(q^{(1)} - q_0^{(1)}), \quad PD-regulator$$
$$(8.29)$$

where: F is the (6×1) vector of external forces and moments acting at the vehicle body MC; $\hat{M}_{22}$, $\hat{M}_{21}$ and $\hat{L}^{(2)}$ are the estimated matrices of the model (8.16); $Q(.)$ is one of the two disposable (3×1) vector functions which determine the character of the function describing the forces/moments in the transient process; $K_F^{(2)}$ and $K_{FI}^{(2)}$ are the (3×3) matrices of the force control gains of the PI or P regulator; $F_0^{(2)}$ is the (3×1) vector of the nominal (programmed) values of forces/moments acting in the considered directions (vertical, roll and pitch); $K_P^{(1)}$ and $K_V^{(1)}$ are the quadratic matrices of the position and velocity control gains of dimension (3×3). These control gains act in directions that are complementary to the directions in which the force/moment control is applied. The feedback control gains included in the dynamic control algorithm are determined using an automatic software procedure established by the practical stability test.

The control block scheme of the vehicle autopilot controller is presented in Figure 8.8. Bearing in mind its practical implementation, it is assumed that the vehicle possesses three active control subsystems on the basis of which it realizes a desired motion in each of six coordinate directions. The control vector τ (Figure 8.8) is partitioned in the two complementary vectors of control signals τ_I and τ_{II}, which compensate for the inaccuracies of motion in the corresponding directions, so that τ_I acts in the z, ϕ and θ direction, and τ_{II} in the x, y and ε direction. The matrices S_1 and S_2 (Figure 8.8) represent the diagonal (6×6) selectivity matrices which serve to separate the control directions. The control signals τ_I and τ_{II} are the reference signals on the lower control level. Control on the executive level depends on the choice of the system's actuators and the concrete solutions of the active systems design. Output variables of the executive control level represent the corresponding signals at the output devices of the system actuators. The vehicle actuators generate the driving forces and moments that can be reduced at the vehicle body MC in the form of the $\tilde{\tau}$ vector (Figure 8.8).

4.4   Synthesis of the supplementary neuro-compensator

A neuro-compensator was added to the synthesized dynamic road vehicle controller. Under the notion of neuro-compensator we assume a compensator structure which is based on the artificial neural network (ANN) with four layers of perceptrons. The proposed model-based controller can ensure stable system motion and the desired quality of its dynamic behaviour if the mathematical model of the system sufficiently exactly describes the system dynamics, as well as if the model parameters deviate relatively little from their real values. In the synthesis of the dynamic controller, we adopted the spatial model of the road vehicle (8.14) which describes relatively well the basic system dynamics in the considered six coordinate directions of motion. However, the mathematical model of the vehicle does not describe fully all the dynamic effects involved in the system during its motion, as for instance elastic modes of the vehicle mechanism, influence of the actuator dynamics, time delay in the driving and control subsystems of the road vehicle, Coulomb's friction in the suspension system joints, some other nonlinearities existing in the system, etc. Measurement errors of the state variables and parameter estimation errors influence the control accuracy of the motion, too. All these phenomena influence the system stability. Thus, in order to meet the requirements of a vehicle controller capable of operatinf under the real exploitation conditions, it is suitable to combine the model-based controller with the knowledge-based nonlinear compensator.

The structure of the hybrid neuro-dynamic controller, whose control scheme is given in Figure 8.8, consists of two functional blocks. The task of the first block is to compensate for the main dynamic effects, while the second block is based on the application of artificial neural network to compensate for an system uncertainties. The automatic control system of an autonomous road vehicle is designed by integrating the dynamic control algorithm and the chosen ANN structure. On the other hand, the neural network can satisfactorily identify the above-mentioned phenomena, which were not comprised by the mathematical model (8.14).

The mentioned identification of non-modelled dynamic effects by ANN is possible by the training procedure of the chosen net structure using some well known learning methods. Here, we used the *standard back propagation learning method*, which is frequently employed to train multilayer nets in similar control tasks [259, 337]. The application of ANN in automatic control of vehicle motion is performed in three phases: (I) the rough off-line training, (II) the fine on-line training and, (III) the implementation of the finally tuned net in the synthesized hybrid controller.

The structure of the learning process for the off-line identification of unmodelled system dynamics is shown in Figure 8.9.

In the beginning, the appropriate input and output signals needed for the network training are selected. For that purpose, manual commands are set for the vehicle which produce variation of the input command signals such as tyre angular velocities ω_i and variation of the ground steering angles δ_i, $i = 1, 4$. Moreover, various external disturbances act upon the vehicle during the motion. These are variations in the road surface profile, changes of intensity and direction of a side wind gust, as well as variation of Coulomb's friction coefficients between the tyre pneumatics and road surface, etc. Using simulation experiments, the global state variables $(q(t), \dot{q}(t), \ddot{q}(t))$ are determined and saved. These state variables serve as input signals to the rough off-line training of the chosen ANN. They are also used for computation of the vehicle

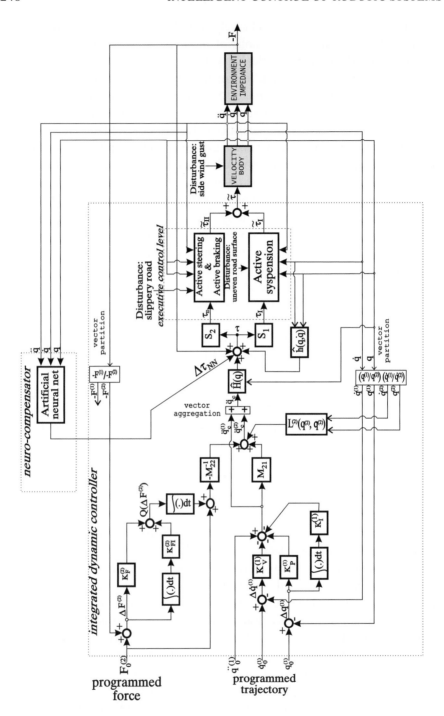

Figure 8.8. Block-scheme of the road vehicle autopilot with centralized dynamic controller and supplementary neuro-compensator

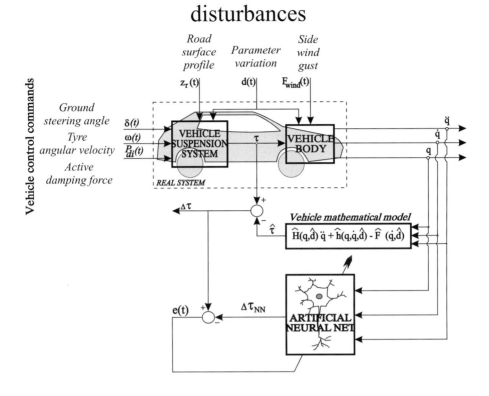

Figure 8.9. Structure of the learning process of the chosen ANN

mathematical model. On the basis of the values of generalized forces $\hat{\tau}(t)$ calculated from the model and their real values $\tau(t)$ obtained by measurement at each time sample on the real system during the ride, the deviation vector $\Delta\tau(t) = \tau(t) - \hat{\tau}(t)$ can be defined. At the same time, the so-called compensation vector of forces/moments $\Delta\tau_{NN}$ is generated at the output layer of the neural network. The values of the (6×1) vector $\Delta\tau_{NN}$ correspond to the values of the identified magnitudes of the forces and torques that are not taken into account by vehicle dynamics modelling. The obtained vectors $\Delta\tau(t)$ and $\Delta\tau_{NN}(t)$ are compared at the discriminator $(e(t) = \Delta\tau(t) - \Delta\tau_{NN}(t))$ (Figure 8.9). The existing error $e(t)$ is used for further tuning of the ANN weighting matrices (Figure 8.9). When the amplitude of the learning error $e(t)$ is below some predetermined value $(e(t) \le E)$, the rough "off-line" learning process is terminated. The next step in the synthesis of the neuro-compensator is the so-called "fine net tuning". The weighting factors in the corresponding matrices of the suitably chosen net architecture, defined by the off-line learning process, serve as the initial values for further "fine tuning" of the neural network. This procedure should be carried out in real time on the real system, in the closed loop (Figure 8.8). When the desired quality of learning process is attained, "fine tuning" is ended.

Here, a suitable multilayer neural net topology for control purposes is proposed. For its training, a standard back propagation algorithm [375, 198] was used. The topology of the proposed net structure intended for control purposes was defined by the four-layer neural network with one input layer, two hidden layers, and one output layer. A sigmoid function was chosen as activation function in the two hidden layers of neurons. Only in the input and output layers was a linear function of identity used as activation function. The output layer permits the net to generate signals from the range $[-1, +1]$. For this reason, at the output of the net, but formally at its input too, the so-called matching gains G_u^i and G_y^i should exist. This is necessary for the realization of a better and faster net convergence, i.e., in order to accelerate the net learning rate.

The proposed neural network has an input layer with $n = 15$ neurons and an output layer with $m = 6$ neurons. Since the relative position of the vehicle MC in the fixed coordinate system connected to a point located on the road surface does not influence the behaviour of the vehicle dynamics, the coordinates $q_1 = x$, $q_2 = y$ and $q_4 = z$ are omitted from the input vector U. Because of that, the vector U should have dimension (15×1):

$$U = [q_3 \ q_5 \ q_6 \ \dot{q}_1 \ \dot{q}_2 \ \dot{q}_3 \ \dot{q}_4 \ \dot{q}_5 \ \dot{q}_6 \ \ddot{q}_1 \ \ddot{q}_2 \ \ddot{q}_3 \ \ddot{q}_4 \ \ddot{q}_5 \ \ddot{q}_6]^T \qquad (8.30)$$

where q_i, $\dot{q}_i$ and $\ddot{q}_i$ are members of the corresponding (6×1) vectors of position q, velocity $\dot{q}$ and vehicle acceleration $\ddot{q}$. The (6×1) output vector Y consists of the compensation forces and torques that should act at the MC of the vehicle body:

$$Y = [\Delta\tau_{NN}^1 \ \Delta\tau_{NN}^2 \ \Delta\tau_{NN}^3 \ \Delta\tau_{NN}^4 \ \Delta\tau_{NN}^5 \ \Delta\tau_{NN}^6]^T \qquad (8.31)$$

where the members $\Delta\tau_{NN}^i$ $(i = 1, 6)$ have characteristics of force (i.e., moment), and they act at the MC of the road vehicle.

Based on simulation experiments, and optimal network topology was defined: the input layer with $n = 15$ neurons, the first hidden layer with $p = 81$ neurons, the second hidden layer with $q = 67$ neurons and, the output layer of the connectionist structure with $m = 6$ neurons. Finally, the experimentally identified values of the matching gains G_u and G_y are defined:

$$G_u^i = 1, \quad \text{for} \ i = 1, 15 \qquad G_y^j = 0.01 \ \text{for} \ j = 1, 6 \qquad (8.32)$$

4.5   Simulation experiments

A characteristic example of vehicle motion along an arbitrary curvilinear trajectory was assumed. The nominal imposed trajectory consisted of two circular ($R_1 = 550$ and $R_2 = 450$ [m] radii) and three linear path segments. In the considered simulation experiment the influence of the following external disturbances on the system stability were imposed: (I) variation of the road surface profile, (II) side wind gust, (III) slippery road, and (IV) time delays in the vehicle actuators. The vehicle motion was simulated against wind so that it had a permanent direction of blowing w.r.t. the longitudinal axis of the path ($\gamma = -60$ [°] clockwise). Force impact intensity was introduced as the variable function $F_w(t) = 2000 \, exp^{-\lambda \Delta t}$ [N]. The "weakness" parameter λ was defined from the condition that the wind force magnitude $F_w(t)$ diminished from 2000 to 0.1 [N] for the time period of $\Delta t = 2$ seconds. Side wind gusts tend to destabilize the system motion and change its forward velocity. The occurrence of a slippery (wet, icy) road causes a variation of the tyre rolling resistance coefficients f_r on the road. They change and they are equal on the same-side tyres ($f_r^1 = f_r^3$ and $f_r^2 = f_r^4$). For a dry road it was assumed that the tyre rolling resistance has the value of $f_r = 0.018$. Variation of the road surface profile was introduced as a "step" function of 8 [mm], appearing on the right pair of tyres. In the considered simulation experiment the effect of the actuator dynamics of the robotized vehicle is taken into account as its time delay $\tau_d = 0.032$ which acts on the system as an internal disturbance. The initial position errors of the vehicle body MC in the longitudinal, lateral and vertical directions respectively, were set as: $\Delta x(0) = -0.25$ [m], $\Delta y(0) = -0.10$ [m], $\Delta z(0) = 0.004$ [m], while the initial angular deviations in the roll, pitch and yaw directions were $\Delta\phi(0) = 0.017$ [rad], $\Delta\theta(0) - 0.017$ [rad] and $\Delta\epsilon(0) = -0.085$ [rad]. The vehicle geometry and dynamic parameters in the considered simulation experiment were taken from [273]. The vehicle behaviour was tested for the synthesized hybrid neuro-dynamic controller with the position/force control algorithm implemented on the tactical control level (Figure 8.8). The control gains were defined so that the system was stable for parameter variation in the range of predetermined limits and for moderate external disturbances. In this simulation experiment, various control commands were simulated in the open loop regime (Figure 8.9) for the purpose of training the chosen neural net structure. Commands on continual variation of the tyre ground steering angles $\delta_i(t)$ ($i = 1, 4$), as well as the command of changing the tyre angular velocities ω_i ($i = 1, 4$), were defined. For the purpose of excitation of the so-called "vertical" vehicle dynamics, variation of the road surface profile z_r (within the limits of $z_r^{max} = \pm 5$ [mm]) upon all wheels was introduced. Time-varying values of the system dynamic parameters such as vehicle integral mass (in the range $\pm 10\%$), inertial moments around the main axis ($\pm 10\%$) as well as coefficients of friction between the tyre pneumatics and the road surface (variation up to 70%), were also simulated. For the net training, changes of the ground steering angles in the form of "step" functions (3.5 and 5.5 [°]) were imposed on the front pair of wheels. On the same pair of wheels, periodical input signals of the same amplitudes and different frequencies of 1 and 1.5 $[Hz]$ were introduced. On the rear wheels, random signals with maximal amplitudes of ± 1.5 [°] were imposed. It was assumed that the vehicle moves at a forward speed $V = 80$ $[km/h]$ with periodic variation of its tyre angular velocities in the range of $\pm 5\%$.

For the neural network training we applied the back propagation method. The network training parameters were preset: the target error $E = 180$, the maximal number of learning epochs $e_{max} = 10^5$ and the initial learning rate of the applied neural network $l_r = 0.01$.

In Figure 8.10, the square criterion during the learning epoch is presented.

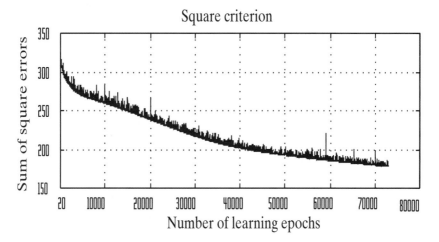

Figure 8.10.   Error in learning process

Also, in Figures 8.11 and 8.12, position and velocity errors during motion with and without the neuro-compensator are shown. The convergence of the learning process when using the neuro-compensator is obvious, as well as the improvement related to tracking errors compared with the case without the neuro-compensator. A satisfactory dynamic vehicle body behaviour in the directions considered was also achieved.

Summarizing on the simulation results it can be concluded that the proposed integrated vehicle controller ensures attaining the main control task - full motion stability. The synthesized vehicle autopilot enables satisfactory dynamic behaviour of the system in its transient regime, as well as a good quality of tracking of the nominal positions of the vehicle body in all six directions of motion. The synthesized hybrid controller also guarantees the realization of the desired payload forces and moments of roll and pitch at the vehicle body MC. The supplementary neuro-compensator gives an additional quality to the automatic control system - a higher robustness to the system modelling inaccuracies. Further improvements should be sought in the application of some other network topology, or in choosing a more efficient method, which would increase the rate of associative learning of the model used in the control. In that way a higher efficiency of the network in the control process should be attained. One of possible directions of further improvements of the proposed controllers should be in the decentralization of the control system, which would additionally simplify the controller design.

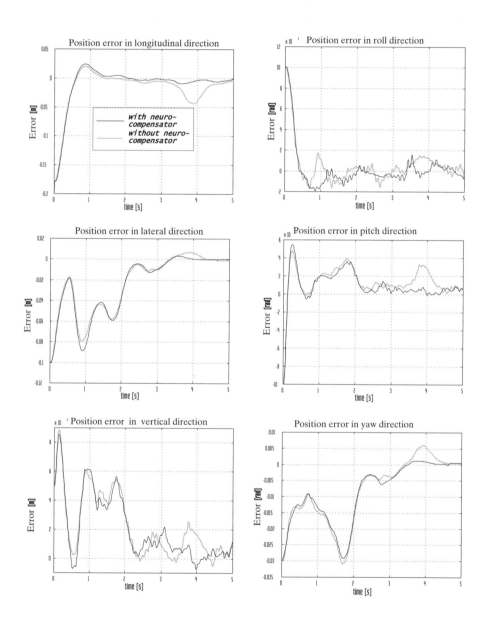

Figure 8.11.   Accuracy indices in position tracking

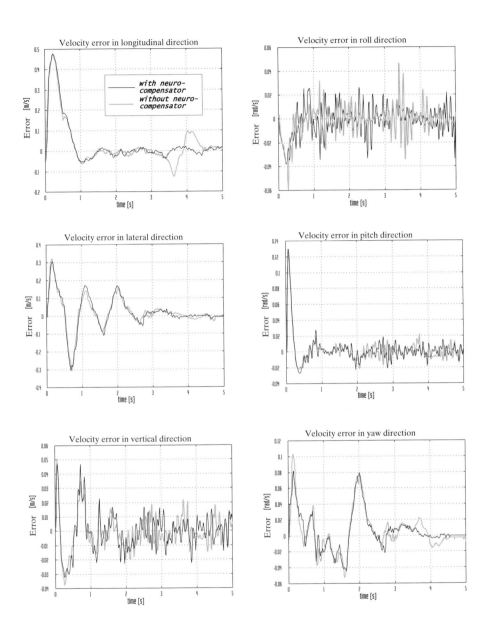

Figure 8.12.   Accuracy indices in velocity tracking

References

[1] M. Aicardi, G. Cannata, and G. Casalino. A learning procedure for position and force control of constrained manipulators. In *Proceedings of the 5th International Conference on Advance Robotics*, pages 423–430, Pisa, June 1991.

[2] M. Aicardi, G. Cannata, and G. Casalino. Hybrid learning control for constrained manipulators. *Advanced Robotics*, 6(1):69–94, 1992.

[3] M-R. Akbarzadeh and M. Jamshidi. Evolutionary fuzzy control of a flexible-link. *Intelligent Automation and Soft Computing*, 3(1):77–88, 1997.

[4] J. Albus. A new approach to manipulator control: The cerebellar model articulation controller. *Journal of Dynamic Systems, Measurement, and Control*, 97:220–227, September 1975.

[5] J. Albus. Outline for a theory of intelligence. *IEEE Transactions on Systems,Man, and Cybernetics*, 21(3):473–509, 1991.

[6] G. Antonelli, S. Chiaverini, and G. Fusco. Experiments of fuzzy real-time path planning for unicycle-like mobile robots under kinematic constraints. In *Proceedings of the IEEE International Conference on Robotcss and Automation*, pages 2147–2152, Washington, D.C., May 2002.

[7] T. Aoki, M. Matsuno, T. Suzuki, and S. Okuma. Motion planning for multiple obstacle avoidance of autonomous mobile robot using hierarchical fuzzy rules. In *Proceedings of the IEEE International Conference on Multisensor Fusion and Integration for Intelligent Systems*, pages 265–271, Las Vegas, 1994.

[8] F. Arai, T. Tanaka, and T. Fukuda. Recurrent neural network modeling and learning control of flexible plates by nonlinear handling system. In *Proceedings of the IEEE International Conference on Robotics and Automation*, pages 2070–2075, San Diego, May 1994.

[9] T. Arakawa and T. Fukuda. Natural motion generation of biped locomotion robot using hierarchical trajectory generation method consisting of GA,EP layers. In *Proceedings of the IEEE International Conference on Robotics and Automation*, pages 211–216, Albuquerque, April 1997.

[10] S. Arimoto and T. Naniwa. Learning control for robot tasks under geometric endpoint constraints. In *Proceedings of the IEEE International Conference on Robotics and Automation*, pages 1914–1919, Nice, May 1992.

[11] R.C. Arkin. *Behaviour-Based Robotics*. MIT Press, Cambridge, 1999.

[12] H. Asada. Teaching and learning of compliance using neural nets: Representation and generalization of nonlinear compliance. In *Proceedings of the IEEE International Conference on Robotics and Automation*, pages 1237–1244, Cincinnati, May 1990.

[13] H. Asada. Representation and learning of nonlinear compliance using neural nets. *IEEE Transactions on Robotics and Automation*, 9(6):863–867, December 1993.

[14] H. Asada and Y. Asari. Direct teaching of tool manipulation skills via the impedance identification of human motions. In *Proceedings of the IEEE International Conference on Robotics and Automation*, pages 1269–1274, Philadelphia, May 1988.

[15] H. Asada and H. Izumi. Direct teaching and automatic program generation for the hybrid control of robot manipulators. In *Proceedings of the IEEE International Conference on Robotics and Automation*, pages 1401–1406, Raleigh, April 1987.

[16] H. Asada and S. Liu. Transfer of human skills to neural net robot controllers. In *Proceedings of the IEEE International Conference on Robotics and Automation*, pages 2442–2448, Sacramento, April 1991.

[17] H. Asada and B-H. Yang. Skill acquisition from human experts through pattern processing of teaching data. In *Proceedings of the IEEE International Conference on Robotics and Automation*, pages 1302–1307, Scottsdale, May 1989.

[18] C.G. Atkeson, A.W. Moore, and S. Schaal. Locally weighted learning. *Artificial Intelligence Review*, 11:11–73, 1997.

[19] A. Bagchi and H. Hartwal. Fuzzy logic based techniques for motion planning of a robot manipulator amongst unknown moving obstacle. *Robotica*, 10:563–573, 1992.

[20] B. Bakker and F. Linaker. Reinforcement learning in partially observable mobile robot domains using unsupervised event extraction. In *Proceedings of the IEEE/RSJ International Conference on Intelligent Robots and Systems*, pages 938–942, Lausanne, October 2002.

[21] T. Balch. *Behavioral Diversity in Learning Robot Teams*. PhD thesis, Georgia Tech, Atlanta, 1998.

[22] J.F. Baldwin and B.W. Pitsworth. Axiomatic approach to implication for approximate reasoning with fuzzy logic. *Fuzzy Sets and Systems*, 3:193–219, 1980.

[23] C. Baroglio, G. Attilio, M. Kaiser, M. Nuttin, and R. Piola. Learning controllers for industrial robots. *Machine Learning*, 1996.

[24] B. Barshan, B. Ayrulu, and S.W. Utete. Neural-network-based target differntiation using sonar for robotics applications. *IEEE Transactions on Robotics and Automation*, 16(4):435–442, August 2000.

[25] A.G. Barto, R.S. Sutton, and C.W. Anderson. Neuron-like adaptive elements that can solve difficult learning control problem. *IEEE Transactions on Systems,Man, and Cybernetics*, 13:834–846, 1983.

[26] D.F. Bassi and G.A. Bekey. Decomposition of neural networks models of robot dynamics: A feasibility study. In W.Webster, editor, *Simulation and AI*, pages 8–13. The Society for Computer Simulation International, 1989.

[27] L. Behera and N. Kirubanandan. A hybrid neural control scheme for visual-motor coordination. *IEEE Control Systems Magazine*, 19(4):34–41, August 1999.

[28] G.A. Bekey and K.Y. Goldberg. *Neural Networks in Robotics*. Kluwer Academic Publishers, Norwell, 1993.

[29] A. Benallegue, B. Daachi, and A.R. Cherif. Stable neural network adaptive control of constrained redundant robot manipulators. In *Proceedings of the IEEE/RSJ International Conference on Intelligent Robots and Systems*, pages 2193–2198, Lausanne, October 2002.

[30] J.G. Bender. An overview of systems studies of automated highway systems. *IEEE Transactions on Vehicular Technology*, 40(1):82–99, 1991.

[31] M. Benreguieg, P. Hoppenot, H. Maaref, E. Colle, and C. Barret. Fuzzy navigation strategy: application to two distinct autonomous mobile robots. *Robotica*, 15(6):609–615, November-December 1997.

[32] A. Billard and M. Mataric. Learning human arm movements by imitation: Evaluation of a biologically-inspired connectionist algorithms. In *Proceedings of the IEEE International Conference on Humanoid Robots HUMANOIDS 2000*, September 2000.

[33] P. Bonissone, P.S. Khedkar, and Y. Chen. Genetic algorithms for automated tuning of fuzzy controllers: A transportation application. In *Proceedings of the Fuzz-IEEE Conference 96*, pages 674–680, New Orleans, 1996.

[34] B. Borovac, D. Katic, and M. Vukobratovic. Connectionist reactive control for robotic assembly tasks by soft sensored grippers. In *Proceedings of the 24th International Symposium on Industrial Robots*, pages 1–8, Tokyo, 1993.

[35] B. Borovac, D. Seslija, and S. Stankovski. Soft sensored grippers in assembly process. In *Proceedings of the IEEE International Conference on Robotics and Automation*, pages 1283–1288, Nice, May 1992.

[36] V. Braitenberg. *Vehicles. Experiments in Synthetic Psychology*. MIT Press, Cambridge, 1984.

[37] R. Brooks. From earwigs to humans. *Robotics and Autonomous Systems*, 20(2-4):291–30, June 1997.

[38] M. Brown and C. Harris, editors. *Neurofuzzy Adaptive Modelling and Control*. Prentice Hall, Englewood Cliffs, 1994.

[39] J.J. Buckley and Y. Hayashi. Fuzzy neural networks: A survey. *Fuzzy Sets and Systems*, 66:1–13, 1994.

[40] J-L. Buessler and J-P. Urban. Visually guided movements: learning with modular neural maps in robotics. *Neural Networks*, 11(7-8):1395–1415, October-November 1998.

[41] G. Capi, Y. Nasu, L. Barolli, M. Yamano, K.Mitobe, and K. Takeda. A neural network implementation of biped robot optimal gait during walking generated by genetic algorithm. In *Proceedings of the 9th Mediterranean Conference on Control and Automation*, Dubrovnik, June 2001.

[42] J.G. Carbonell. Introduction: Paradigms for machine learning. *Artificial Intelligence*, (40):1–9, 1989.

[43] R. Carelli, F. Camacho, and D. Patino. A neural network based feedforward adaptive controller for robots. *IEEE Transactions on Systems, Man, and Cybernetics*, 25(9):1281–1288, 1995.

[44] G.A. Carpenter and S. Grossberg. The ART of adaptive pattern recognition by a self-organizing neural network. *Computer*, pages 77–88, Marth 1988.

[45] J. Carusone and G.M.T. D'Eleuterio. The feature 'CMAC': A neural-network-based vision system for robotic control. In *Proceedings of the IEEE International Conference on Robotics and Automation*, pages 2959–2964, Leuven, May 1998.

[46] D.H. Cha and H.S. Cho. A neurofuzzy model-based compliance controller with application to a telerobot system. *Control Engineering Practice*, 4(3):319–330, May 1996.

[47] C.L.P. Chen and A.D. McAulay. Robot kinematics learning learning computations using polynomial neural networks. In *Proceedings of the IEEE International Conference on Robotics and Automation*, pages 2638–2643, Sacramento, May 1991.

[48] M. Chen and A.M.S. Zalzala. A genetic approach to motion planning of redundant mobile manipulator systems considering safety and optimization. *Journal of Robotic Systems*, 14(7):529–544, 1997.

[49] M-Y. Cheng and C-S.Lin. Genetic algorithm for control design of biped locomotion. *Journal of Robotic Systems*, 14(5):365–373, May 1997.

[50] K.C. Cheok, G-E. Smid, K. Kobayashi, J.L. Overholt, and P. Lescoe. A fuzzy logic intelligent control system paradigm for an in-line-of-sight leader-following HMMWV. *Journal of Robotic Systems*, 14(6):407–419, 1997.

[51] C.K. Chui. *An Introduction to Wavelets*. Academic Press, London, 1992.

[52] D. Cliff and G. F. Miller. Tracking the red queen: Measurements of adaptive progress in co-evolutionary simulations. In F. Morgan, A. Moreno, J. J. Merelo, and P. Chacton, editors, *Advances in Artifcial Life: Proceedings of the Third European Conference on Artificial Life*. Springer Verlag, Berlin, 1995.

[53] C.A. Coello Coello, A.D. Christiansen, and A.H. Aguirre. Using a new GA-based multi-objective optimization techniques for the design of robot arms. *Robotica*, 16(4):401–414, July-August 1998.

[54] M. Cohen and T. Flash. Learning impedance parameters for robot control using an associative search network. *IEEE Transactions on Robotics and Automation*, 7(3):382–390, June 1991.

[55] S. Commuri, S. Jagannathan, and F.L. Lewis. CMAC neural network control of robot manipulators. *Journal of Robotic Systems*, 14(6):466–482, June 1997.

[56] M. Dapper, R. Maas, V. Zahn, and R. Eckmiller. Neural force control (NFC) applied to industrial manipulators in interaction with moving rigid objects. In *Proceedings of the IEEE International Conference on Robotics and Automation*, pages 2048–2053, Leuven, May 1998.

[57] Y. Davidor. A genetic algorithm applied to robot trajectory generation. In L.Davis, editor, *Handbook of Genetic Algorithms*, pages 144–165. Van Nostrand Reinhold, New York, 1991.

[58] R. De Guiseppe, F. Taurisano, C. Distante, and A. Anglani. Visual servoing of a robotic manipulator based on fuzzy logic control. In *Proceedings of the IEEE International Conference on Robotics and Automation*, pages 1487–1494, Detroit, May 1999.

[59] A. De Luca and C. Manes. On the modelling of robots in contact with dynamic environments. In *Proceedings of the 5-th International Conference on Advanced Robotics*, pages 568–574, Pisa, 1991.

[60] A. De Luca and C. Manes. Modelling of robots in contact with the dynamic environment. *IEEE Transactions on Robotics and Automation*, 10(4):542–548, 1994.

[61] C.W. De Silva and A.G.J. MacFarlane. *Knowledge-Based Control with Application to Robots*. Springer Verlag, Berlin, 1989.

[62] C.W. De Silva and A.G.J. McFarlane. Knowledge-based control approach for robotic manipulator. *International Journal of Control*, 50(1):249–273, 1989.

[63] D.R. Deyong, J. Polson, R. Moore, C. Weng, and J. Lara. Fuzzy and adaptive control simulation for a walking machine. *IEEE Control Systems Magazine*, 12(3):43–50, 1992.

[64] H. Ding and J. Wang. Recurrent neural networks for minimum infinity-norm kinematic control of redundant manipulators. *IEEE Transactions on Systems,Man, and Cybernetics, Part A*, 29:269–276, 1999.

[65] D.R. Dodds. Fuzzines in knowledge-based robotic systems. *Fuzzy Sets and Systems*, 26(2):179–193, 1988.

[66] P.I. Doerschuk, W.E. Simon, V. Nguyen, and A. Li. A modular approach to intelligent control of a simulated jointed leg. *IEEE Robotics & Automation Magazine*, 5(2):12–21, June 1998.

[67] L. Doitsidis, K.P. Valavanis, and N.C. Tsourveloudis. Fuzzy logic based autonomous skid steering vehicle navigation. In *Proceedings of the IEEE International Conference on Robotcss and Automation*, pages 2171–2176, Washington, D.C., May 2002.

[68] K. Doya, H. Kimura, and M. Kawato. A modular approach to intelligent control of a simulated jointed leg. *IEEE Control Systens Magazine*, 21(4):42–54, August 2001.

[69] D. Dubois and H. Prade. Fuzzy logics and the generalized modus ponens revisited. *Cybernetic Systems*, 15:3–4, 1984.

[70] P. Dyer and S.R. McReynolds. *The computation and theory of optimal control*. Academic Press, New York, 1970.

[71] K. Endo, F. Yamasaki, T. Maeno, and H. Kitano. A method for co-evolving morphology and walking pattern of biped humanoid robot. In *Proceedings of the IEEE International Conference on Robotics and Automation*, pages 1950–1956, Washington, D.C., May 2002.

[72] A.M. Erkmen and M. Durna. Genetic algorithm-based optimal regrasping with the anthrobot 5-fingered robot hand. In *Proceedings of the IEEE International Conference on Robotics and Automation*, pages 3329–3334, Leuven, May 1998.

[73] K. Feng and L.L. Hoberock. A modified ARTMAP network, with aopplications to scheduling of a robot-vision-tracking system. *Journal of Dynamic Systems, Measurement and Control*, 118:1–8, March 1996.

[74] J.J. Fernandez and I.D. Walker. Biologically inspired robot grasping using genetic programming. In *Proceedings of the IEEE International Conference on Robotics and Automation*, pages 3032–3039, Leuven, May 1998.

[75] FETRO. www.cordis.lu/ist/fetro.htm, 2002.

[76] D. Floreano. Emergence of home-based foraging strategies in ecosystems of neural networks. In J. Meyer, H. L. Roitblat, and S. W. Wilson, editors, *Proceedings of the Second International Conference on Simulation of Adaptive Behavior*. MIT Press - Bradford Books, Cambridge, 1993.

[77] D. Floreano. Evolutionary robotics in artificial life and behavior engineering. In T. Gomi, editor, *Proceedings of the Second International Conference on Simulation of Adaptive Behavior*. AAI Books, Ontario, 1998.

[78] D. Floreano and F. Mondada. Evolution of homing navigation in a real mobile robot. *IEEE Transactions on Systems, Man, and Cybernetics*, 26(3):396–407, June 1996.

[79] D. Floreano and F. Mondada. Evolutionary neurocontrollers for autonomous mobile robots. *Neural Networks*, 11:1461–1478, 1998.

[80] D. Floreano and S. Nolfi. God save the red queen! Competition in coevolutionary robotics. In J. Koza, K. Deb, M. Dorigo, D. Fogel, M. Garzon, H. Iba, and R. L. Riolo, editors, *Proceedings of the 2nd International Conference on Genetic Programming*. Morgan Kaufmann, San Mateo, 1997.

[81] D. Floreano, S. Nolfi, and F. Mondada. Competitive co-evolutionary robotics: From theory to practice. In *From Animals to Animats 5*, pages 515–524. The MIT Press, Cambridge,MA, 1998.

[82] K.S. Fu. Learning control systems and intelligent control systems: An intersection of artificial intelligence and automatic control. *IEEE Transactions on Automatic Control*, 16:70–72, 1971.

[83] Y. Fu, R.Sharma, and M. Zeller. Vision-based motion planning for a robot arm using topology representing networks. In *Proceedings of the IEEE International Conference on Robotics and Automation*, pages 1900–1905, San Francisko, April 2000.

[84] S. Fuchun, S. Zengqi, and S. Rongjun. Stable adaptive control for robot trajectory tracking using neural network. In *Proceedings of the IEEE International Conference on Robotics and Automation*, pages 3440–3445, Minneapolis, April 1996.

[85] S. Fukami, M. Mizumoto, and K. Tanaka. Some considerations of fuzzy conditional inference. *Fuzzy Sets and Systems*, 4:243–273, 1980.

[86] T. Fukuda, Y. Komata, and T. Arakawa. Stabilization control of biped locomotion robot based learning with GAs having self-adaptive mutation and recurrent neural network. In *Proceedings of the IEEE International Conference on Robotics and Automation*, pages 217–222, Albuquerque, April 1997.

[87] T. Fukuda, K. Mase, and Y. Hasegawa. Robot hand manipulation by evolutionary programming. In *Proceedings of the IEEE International Conference on Robotics and Automation*, pages 2458–2463, Detroit, May 1999.

[88] T. Fukuda, T. Shibata, M. Tokita, and T. Mitsuoka. Adaptation and learning by neural network for robotic manipulator. In *Proceedings of the IMACS International Symposium on Mathematical and Intelligent Models in System Simulation*, Brussels, September 1990.

[89] J. Gallagher, R. Beer, K. Espenschiel, and R. Ouinn. Application of evolved locomotion controllers to a hexapod robot. *Robot.Autonomous Syst.*, 19(1):95–103, 1996.

[90] L-M. Garcia, A.A.F. Oliveira, R.A. Grupen, D.S. Wheeler, and A.H. Fagg. Tracing patterns and attention: Humanoid robot cognition. *IEEE Intelligent Systems*, pages 70–77, July/August 2000.

[91] S.S. Ge, T.H. Lee, and C.J. Harris. *Adaptive Neural Network Control of Robotic Manipulators*. World Scientific, Singapore, 1998.

[92] S.S. Ge, T.H. Lee, and G. Zhu. Genetic algorithm tuning of Lyapunov-based controllers: An application to a single-link flexible robot system. *IEEE Transactions on Industrial Electronics*, 43(5):567–574, October 1996.

[93] S.F. Giszter, K.A. Moxon, I.A. Rybak, and J.K. Chaplin. Neurobiological and neuro-robotic approaches to a control architecture for a humanoid motor. *IEEE Intelligent Systems*, pages 64–69, July/August 2000.

[94] S.J. Go and M.C. Lee. Fuzzy-sliding mode control with the self tuning fuzzy inference based on genetic algorithm. In *Proceedings of the IEEE International Conference on Robotics and Automation*, pages 2124–2129, San Francisko, April 2000.

[95] J. Godjevac. *Neuro-Fuzzy Controllers*. Presses Polytechniques et Universitaires Romandes, Lausanne, 1997.

[96] P. Goel, G. Dedeoglu, S.I. Roumeliotis, and G.S. Sukhatme. Fault detection and identification in a mobile robot using multiple estimation and neural network. In *Proceedings of the IEEE International Conference on Robotics and Automation*, pages 2302–2309, San Francisko, April 2000.

[97] D.E. Goldberg. *Genetic Algorithms in Search,Optimization,and Machine Learning*. Addison Wesley, Reading, 1989.

[98] T. Gomi and K. Ide. Emergence of gaits of a legged robot by collaboration through evolution. In *Proceedings of the International Symposium on Artificial Life and Robotics*. Springer-Verlag, Berlin,Germany, 1997.

[99]   S. Goodridge and R. Luo. Fuzzy behaviour fusion for reactive control of an autonomous mobile robot: MARGE. In *Proceedings of the IEEE International Conference on Robotics and Automation*, pages 1622–1627, San Diego, May 1994.

[100] D. Gracanin, K.P. Valavanis, N.C. Tsourveloudis, and M. Matijasevic. Virtual-environment-based navigation and control of underwater vehicles. *IEEE Robotics & Automation Magazine*, 6(2):53–62, June 1999.

[101] W. Gueaieb, F. Karray, S. Al-Sharhan, and O. Basir. A hybrid adaptive fuzzy approach for the control of cooperative manipulators. In *Proceedings of the IEEE International Conference on Robotcss and Automation*, pages 2153–2158, Washington, D.C., May 2002.

[102] A. Guez and J. Selinsky. Neurocontroller design via supervised and unsupervised leaning. *Journal of Intelligent and Robotic systems*, 2(2-3):307–335, 1989.

[103] M. Guihard and P. Gorce. Dynamic control of a biomechanical inspired biped:BIPMAN. In *Proceedings of the 4th International Conference on Climbing and Walking Robots*, Karlsruhe, September 2001.

[104] V. Gullapali, A. Barto, and R. Grupen. Learning admittance mappings for force-guided assembly. In *Proceedings of the IEEE International Conference on Robotics and Automation*, pages 2633–2638, San Diego, May 1994.

[105] V. Gullapali, R. Grupen, and A. Barto. Learning reactive admittance control. In *Proceedings of the IEEE International Conference on Robotics and Automation*, pages 1475–1481, Nice, May 1992.

[106] V. Gullapalli. A stochastic reinforcement learning algorithm for learning real-valued functions. *Neural Networks*, 3:671–692, 1990.

[107] J. Guo and V. Cherkasky. A solution to the inverse kinematic problem in robotics using neural network processing. In *Proceedings of the IEEE International Joint Conference on Neural Networks*, pages 299–304, Washington, 1989.

[108] Q. Guo. An adaptive robust compensation control schemes using ANN for a redundant robot manipulator in the task space. In *Proceedings of the IEEE International Conference on Robotics and Automation*, pages 1908–1913, Nice, May 1992.

[109] L.B. Gutierrez, F.L. Lewis, and J.L. Lowe. Implementation of a neural network tracking controller for a single flexible links: Comparisom with PD and PID controllers. *IEEE Transactions on Industrial Electronics*, 45(2):307–318, April 1998.

[110] H. Hagras, V. Callaghan, M. Colley, and M. Carr-West. A fuzzy-genetic based embedded-agent approach to learning and control in agricultural autonomous vehicles. In *Proceedings of the IEEE International Conference on Robotics and Automation*, pages 1005–1010, Detroit, May 1999.

[111] K.A. Han-Gyoo. A new interpretation of force/servo control of robot arms. In *Proceedings of the IEEE Symposium on Intelligent Robots and Systems*, pages 1623–1627, 1991.

[112] I. Harvey, P. Husbands, and D. Cliff. Seeing the light: Artificial evolution,real vision. In D. Cliff, P. Husbands, J. Meyer, and S. W. Wilson, editors, *From Animals to Animats III:*

Proceedings of the Third International Conference on Simulation of Adaptive Behavior. MIT Press-Bradford Books, Cambridge, 1994.

[113] H. Hashimoto, T. Kubota, M. Sato, and F. Harashima. Visual control of robotic manipulator based on neural networks. *IEEE Transactions on Industrial Electronics*, 39(6):490–496, 1992.

[114] R. Haupt and S.E. Haupt, editors. *Practical Genetic Algorithms.* John Wiley & Sons, New York, 1998.

[115] S. Haykin. *Neural Networks - A Comprehensive foundation*, volume 1-2. Mackmillan Publishing Company, Englewood Cliffs, 1994.

[116] J.K. Hedrick, M. Tomizuka, and P. Varaiya. Control issues in automated highway systems. *IEEE Control Systems Magazine*, 14(6):21–32, 1994.

[117] K. Hirota, Y. Arai, and S. Hachisu. Fuzzy control robot arm playing two-dimensional ping-pong game. *Fuzzy Sets and Systems*, 32(2):149–159, 1989.

[118] F. Hlawatsch and G.F. Boudreaux-Bartels. Linear and quadratic time-frequency signal representations. *IEEE Signal Processing Magazine*, 11(2):21–67, 1992.

[119] C. Hocaoglu and A.C. Sanderson. Multi-dimensional path planning using evolutionary computation. In *Proceedings of the IEEE World Congress on Computational Intelligence*, pages 165–170, Anchorage, May 1998.

[120] N.E. Hodge and M.B. Trabia. Steering fuzzy logic controller for an autonomous vehicle. In *Proceedings of the IEEE International Conference on Robotics and Automation*, pages 2482–2488, Detroit, May 1999.

[121] N. Hogan. Impedance control: An approach to manipulation, part 1.- theory, part 2.- implementation, part 3.- application. *ASME Journal of Dynamic Systems, Measurement and Control*, 107:1–24, 1985.

[122] J.H. Holland. *Adaptation in Natural and Artificial Systems.* MIT Press, Cambridge, 1975.

[123] A. Homaifar, M. Bikdash, and V. Gopalan. Design using genetic algorithms of hierarchical hybrid fuzzy-PID controllers of two-link robotic arms. *Journal of Robotic Systems*, 14(5):449–463, June 1997.

[124] J. Hopfield. Neural networks and physical systems with emergent collectional computational abilities. *Proceedings of the National Academy of Science*, pages 2554–2558, April 1988.

[125] S. Horikawa, T. Furuhashi, and Y. Uchikawa. On identification of structures in premises of a fuzzy model using a fuzzy neural network. In *Proceedings of the 2nd IEEE International Conference on Fuzzy Systems*, pages 661–666, 1993.

[126] B. Horne, M. Jamshidi, and N. Vadiee. Neural networks in robotics: A survey. *Journal of Intelligent and Robotic Systems*, 3:51–66, 1990.

[127] F-Y. Hsu and L-C. Fu. New design of adaptive fuzzy hybrid force/position control for robot manipulators. In *Proceedings of the IEEE International Conference on Robotics and Automation*, pages 863–868, Nagoya, May 1995.

[128] F-Y. Hsu and L-C. Fu. Adaptive fuzzy hybrid force/position control for robot manipulators following contours of an uncertain object. In *Proceedings of the IEEE International Conference on Robotics and Automation*, pages 2232–2237, Minneapolis, April 1996.

[129] F-Y. Hsu and L-C. Fu. A new adaptive fuzzy hybrid force/position control for intelligent robot deburring. In *Proceedings of the IEEE International Conference on Robotics and Automation*, pages 2476–2481, Detroit, May 1999.

[130] F-Y. Hsu and L-C. Fu. Intelligent robot deburring using adapive fuzzy hybrid position / force control. *IEEE Transactions on Robotics and Automation*, 16(4):325–335, August 2000.

[131] J. Hu, J. Pratt, and G. Pratt. Stable adaptive control of a bipedal walking robot with CMAC neural networks. In *Proceedings of the IEEE International Conference on Robotics and Automation*, pages 1950–1956, Detroit, May 1999.

[132] S. Hu, M. Ang Jr, and H. Krishnan. Neural network controller for constrained robot manipulators. In *Proceedings of the IEEE International Conference on Robotics and Automation*, pages 1906–1911, San Francisko, April 2000.

[133] T. Hu, S.X. Yang, F. Wang, and G.S. Mittal. A neural network controller for a non-holonomic mobile robot with unknown robot parameters. In *Proceedings of the IEEE International Conference on Robotcss and Automation*, pages 3540–3545, Washington, D.C., May 2002.

[134] S-J. Huang and R-J. Lian. A hybrid fuzzy logic and neural network algorithm for robot motion control. *IEEE Transactions on Industrial Electronics*, 44(3):408–417, June 1997.

[135] S.H. Huang and H-C. Zhang. Artificial neural networks in manufacturing:concepts, applications and perspectives. *IEEE Transactions on Components,Packaging, and Manufacturing Technology - Part A*, 17(2):212–228, June 1994.

[136] M-C. Hwang and X. Hu. A robust position/force learning controller of manipulators via nonlinear h control and neural networks. *IEEE Transactions on Systems, Man, and Cybernetics, Part B: Cybernetics*, 30(2):310–321, April 2000.

[137] S. Inagaki, I. Kshiro, and M. Yamamoto. Analysis of vehicle stability in critical cornering using phase-plane method. In *Proceedings of International Symposium on Advanced Vehicle Control*, pages 287–292, Tsukuba, 1994.

[138] S. Irmscher, E. Hees, and T .Kutsche. A controlled suspension system with continuously adjustable damping force. In *Proceedings of International Symposium on Advanced Vehicle Control*, pages 325–330, Tsukuba, 1994.

[139] H. Ishigami, T. Fukuda, T. Shibata, and F. Arai. Structure optimization of fuzzy neyral network by genetic algorithm. *Fuzzy Sets and Systems*, 71:257–264, 1995.

[140] K. Ishii, T. Fujii, and T. Ura. Neural network system for on-line controller adaptation and its application to underwater robot. In *Proceedings of the IEEE International Conference on Robotics and Automation*, pages 756–761, Leuven, May 1998.

[141] S. Ishikawa. A method of indoor mobile robot navigation by using fuzzy control. In *Proceedings of the IEEE Symposium on Intelligent Robots and Systems*, pages 1013–1018, 1991.

[142] C. Isik and A.M. Meystel. Pilot level of a hierarchical controller for an unamnned mobile robot. *IEEE Journal of Robotics and Automation*, 4(3):241–255, 1988.

[143] M. Isogai, F. Arai, and T. Fukuda. Modeling and vibration control with neural network for flexible multi-link structures. In *Proceedings of the IEEE International Conference on Robotics and Automation*, pages 1096–1101, Detroit, May 1999.

[144] K. Ito and F. Matsuno. A study of reinforcement learning for the robot with many degrees of freedom - acquisition of locomotion parameters for multi legged robot. In *Proceedings of the IEEE International Conference on Robotics and Automation*, pages 3392–3398, Washington, D.C., May 2002.

[145] A.G. Ivakhenko. The group method of data handling in prediction problems. *Soviet Automatic Control*, 9(6):21–30, 1976.

[146] M. Ivanescu, A.M. Popescu, and D. Popescu. Interception for a walking robot by fuzzy observer and fuzzy controller. In *Proceedings of the 4th International Conference on Climbing and Walking Robots*, Karlsruhe, September 2001.

[147] W. Jacak, editor. *Intelligent Robotic Systems - Design, Planning and Control*. Kluwer Academic Publishers, Norwell, 1999.

[148] N. Jacobi. Half-baked, ad-hoc and noisy: Minimal simulations for evolutionary robotics. In P.Husbands and I.Harvey, editors, *Proceedings of the 4th European Conference on Artificial Life*. MIT Press, Cambridge, 1997.

[149] S. Jagannathan and F.L. Lewis. Discrete-time neural net controller with quarenteed performance. In *Proceedings of the American Control Conference*, volume 3, pages 3334–3339, 1994.

[150] N. Jakobi, P. Husbands, and I. Harvey. Noise and the reality gap: The use of simulation in evolutionary robotics. In J.Moran, J.Merelo, and P.Chancon, editors, *Advances in Artificial Life: Proceedings of the Third European Conference on Artificial Life*. Springer-Verlag, Berlin,Germany, 1995.

[151] J-S.R. Jang. ANFIS, adaptive-network-based fuzzy inference systems. *IEEE Transactions on Systems,Man, and Cybernetics*, 23(3):665–685, 1992.

[152] D. Jeon and M. Tomizuka. Learning hybrid force and position control of robot manipulators. In *Proceedings of the IEEE International Conference on Robotics and Automation*, pages 1455–1460, Nice, May 1992.

[153] X. Jin. Decentralized adaptive fuzzy control of robot manipulators. *IEEE Transactions on Systems,Man, and Cybernetics, Part B: Cybernetics*, 28(1):47–57, 1998.

[154] Y. Jin, A.G. Pipe, and A. Winfield. Robot trajectory control using neural network - theory and PUMA simulations. In *Proceedings of the IEEE International Symposium on Intelligent Control*, pages 370–375, 1994.

[155] M.I. Jordan. Computational aspects of motor control and motor learning. In H. Heuer and S.W.Keele, editors, *Handbook of perception and action*. Academis Press, New York, 1996.

[156] J-G. Juang. Fuzzy neural network approaches for robotic gait synthesis. *IEEE Transactions on Systems,Man, and Cybernetics, Part B: Cybernetics*, 30(4):594–601, 2000.

[157] J-G. Juang and C-S. Lin. Gait synthesis of a biped robot using backpropagation through time algorithm. In *Proceedings of the IEEE International Joint Conference on Neural Networks*, pages 1710–1715, Washington,D.C., 1996.

[158] L. Kaelbling, M.Littman, and A. Moore. Reinforcement learning : A survey. *Journal of AI*, 4:237–285, 1996.

[159] M. Kaiser and R. Dillmann. Building elementary robot skills from human demonstration. In *Proceedings of the IEEE International Conference on Robotics and Automation*, pages 2700–2705, Minneapolis, April 1996.

[160] T. Kamegawa, F. Matsuno, and R. Chatterjee. Proposition of twisting mode of locomotion and GA based motion planning for transition of locomotion modes of 3-dimensional snake-like robot. In *Proceedings of the IEEE International Conference on Robotics and Automation*, pages 1507–1512, Washington, D.C., May 2002.

[161] B. Karan and M. Vukobratovic. Design of a non-adaptive fuzzy logic-based gain tuning scheme for enhancement of free-space robot motion control. In *Proceedings of the 1st ECPD International Conference on Advanced Robotics and Intelligent Automation*, pages 105–111, Athens, 1995.

[162] B. Karan and M. Vukobratovic. Application of fuzzy logic to simultaneous dynamic position-force control of robotic manipulators. In *Proceedings of the 2nd ECPD International Conference on Advanced Robotics,Intelligent Automation and Active Systems*, pages 80–85, Vienna, 1996.

[163] D. Katic. Genetic algorithm tuning of connectionist controller for compliant robotic tasks. In *Proceedings of the 14th IFAC World Congress*, Beijing, July 1999.

[164] D. Katic and S. Stankovic. Fast learning algorithms for training of multilayer perceptrons based on Extended Kalman Filter. In *Proceedings of the IEEE International Conference on Neural networks*, pages 435–440, Washington,DC, June 1996.

[165] D. Katic and M. Vukobratovic. Decomposed connectionist architecture for fast and robust learning of robot dynamics. In *Proceedings of the IEEE International Conference on Robotics and Automation*, pages 2064–2069, Nice, May 1992.

[166] D. Katic and M. Vukobratovic. Connectionist approaches to the control of manipulation robots at the executive hierarchical level: An overview. *Journal of Intelligent and Robotic Systems*, 10:1–36, 1994.

[167] D. Katic and M. Vukobratovic. Highly efficient intelligent learning control for manipulation robots by feedforward neural networks. In *Preprints of the 7th IFAC/IFORS/IMACS Symposium on Large Scale Systems:Theory and Applications*, pages 779–784, London, July 1995.

[168] D. Katic and M. Vukobratovic. Highly efficient robot dynamics learning by decomposed connectionist feedforward control structure. *IEEE Transactions on Systems,Man, and Cybernetics*, 25(1):145–158, 1995.

[169] D. Katic and M. Vukobratovic. The application of connectionist structures for learning impedance control in robotic contact tasks. *Applied Intelligence*, 7(4):315–326, November 1997.

[170] D. Katic and M. Vukobratovic. Robot compliance algorithm based on neural network classification and learning of robot-environment dynamic models. In *Proceedings of the IEEE International Conference on Robotics and Automation*, pages 2632–2637, Albuquerque, April 1997.

[171] D. Katic and M. Vukobratovic. A neural network-based classification of environment dynamic models for compliant control of manipulation robots. *IEEE Transactions on Systems, Man, and Cybernetics, Part B: Cybernetics*, 28(1):58–69, February 1998.

[172] M. Kawato. Internal models for motor control and trajectory planning. *Current Opinion in Neurobiology*, 12:718–727, 1999.

[173] M. Kawato, Y. Uno, R. Isobe, and R. Suzuki. Hierarchical neural network model for voluntary movement with application to robotics. *IEEE Control Systems Magazine*, 57:169–185, 1987.

[174] R. Kelly, R. Carelli, M. Amestegui, and R. Ortega. On adaptive impedance control of robot manipulators. In *Proceedings of the IEEE International Conference on Robotics and Automation*, pages 572–577, Scottsdale, May 1989.

[175] C. Kemal. Fuzzy rule-based motion controller for an autonomous mobile robot. *Robotica*, 7(1):37–42, 1989.

[176] K. Kiguchi and T. Fukuda. Robot manipulator contact force control application of fuzzy - neural network. In *Proceedings of the IEEE International Conference on Robotics and Automation*, pages 875–880, Nagoya, May 1995.

[177] K. Kiguchi and T. Fukuda. Fuzzy neural friction compensation method of robot manipulation during position / force control. In *Proceedings of the IEEE International Conference on Robotics and Automation*, pages 372–377, Minneapolis, April 1996.

[178] K. Kiguchi and T. Fukuda. Intelligent position / force controller for industrial robot manipulators - application of fuzzy neural networks. *IEEE Transactions on Industrial Electronics*, 44(6):753–761, December 1997.

[179] K. Kiguchi and T. Fukuda. Robot manipulator hybrid control for an unknown environment using visco-elastic neural networks. In *Proceedings of the IEEE International Conference on Robotics and Automation*, pages 1447–1452, Leuven, May 1998.

[180] K. Kiguchi and T. Fukuda. Fuzzy selection of fuzzy-neuro robot force controllers in an unknown environment. In *Proceedings of the IEEE International Conference on Robotics and Automation*, pages 1182–1187, Detroit, May 1999.

[181] K. Kiguchi, K. Watanabe, K. Izumi, and T. Fukuda. Application of multiple fuzzy-neuro force controllers in an unknown environment using genetic algorithms. In *Proceedings of the IEEE International Conference on Robotics and Automation*, pages 2106–2111, San Francisco, April 2000.

[182] H. Kim and P.I. Ro. Robust learning control by neural network for an active four wheel steering system. *Intelligent Automation and Soft Computing*, 2(3):1–14, 1996.

[183] K-Y. Kim and Y-S. Yoon. Practical inverse kinematics of a kinematically redundant robot using a neural network. *Advanced Robotics*, 6(4):431–440, 1992.

[184] S. Kim and W.R. Hamel. Force assistance function for human machine cooperative telerobotics using fuzzy logic. In *Proceedings of the IEEE International Conference on Robotcss and Automation*, pages 2165–2170, Washington, D.C., May 2002.

[185] Y.H. Kim and F.L. Lewis. Optimal design of CMAC neural-network controller for robot manipulators. *IEEE Transactions on Systems,Man, and Cybernetics, Part C*, 30(1):22–31, 2000.

[186] P. Kiriazov. Learning robots to move: Biological control concepts. In *Proceedings of the 4th International Conference on Climbing and Walking Robots*, Karlsruhe, September 2001.

[187] S. Kitamura, Y. Kurematsu, and Y. Nakai. Application of the neural network for the trajectory planning of a biped locomotion robot. *Neural Networks*, 1:344–356, 1988.

[188] H. Kobayashi, K. Uchida, and R. Matsuzaki. Robot vision system with self-learning mechanism. *Journal of Artificial Neural Networks*, 2(1-2):137–144, 1995.

[189] R. Koeppe, A. Breidenbach, and G. Hirzinger. Skill representation and acquisition of compliant motions using a teach device. In *Proceedings of the IEEE/RSJ International Conference on Intelligent Robots and Systems*, pages 1–8, Osaka, November 1996.

[190] T. Kohonen. Self-organized formation of topology correct feature maps. *Biological Cybernetics*, (43):59–69, 1982.

[191] T. Kondo and K. Ito. A reinforcement learning with adaptive state space recruitment strategy for real autonomous mobile robots. In *Proceedings of the IEEE/RSJ International Conference on Intelligent Robots and Systems*, pages 897–902, Lausanne, October 2002.

[192] N. Kubota, T. Fukuda, and K. Shimojama. Trajectory planning of cellular manipulator system using virus-evolutionary genetic algorithm. *Robotics and Autonomous Systems*, 19:85–94, 1996.

[193] G.M. Kulali, M. Gevher, A.M. Erkmen, and I. Erkmen. Intelligent gait synthesizer for serpentine robots. In *Proceedings of the IEEE International Conference on Robotcss and Automation*, pages 1513–1518, Washington, D.C., May 2002.

[194] A.L. Kun and W. Miller III. Control of variable - speed gaits for a biped robot. *IEEE Robotics and Automation Magazine*, 6(3):19–29, September 1999.

[195] S-Y. Kung and J-N. Hwang. Neural network architectures for robotic applications. *IEEE Transactions on Robotics and Automation*, 5(5):641–657, October 1989.

[196] M. Kuperstain. Adaptive visual-motor coordination in multijoint robots using parallel architecture. In *Proceedings of the IEEE International Conference on Robotics and Automation*, pages 1595–1601, Raleigh, March 1987.

[197] Y. Kurematsu, O. Katayama, M. Iwata, and S. Kitamura. Autonomous trajectory generation of a biped locomotive robot. In *Proceedings of the IEEE International Joint Conference on Neural Networks*, pages 1983–1988, 1991.

[198] J.G. Kuschewski, S. Hui, and S.H. Zak. Application of feedforward neural network to dynamical system identification and control. *IEEE Transactions on Control Systems Technology*, 1(1):37–49, March 1993.

[199] J.M. Won Kwon, Y.D. and J.S. Lee. Control of mobile robot by using evolutionary fuzzy controller. In *Proceedings of the IEEE International Conference on Robotics and Automation*, pages 422–427, Detroit, May 1999.

[200] F. Lange. A learning concept for improving robot force control. In *Proceedings of the IFAC Symposium on Robot Control*, pages 79.1–79.6, Karlsruhe, October 1988.

[201] F. Lange and G. Hirzinger. Iterative self-improvement of force feedback control in contour tracking. In *Proceedings of the IEEE International Conference on Robotics and Automation*, pages 1399–1404, Nice, May 1992.

[202] D.A. Lawrence. Impedance control stability properties in common implementations. In *Proceedings of the IEEE International Conference on Robotics and Automation*, pages 1185–1190, Philadelphia, April 1988.

[203] M.A. Johnson Leahy, M.B. and S.K. Rogers. Neural network payload estimation for adaptive robot control. *IEEE Transactions on Neural Networks*, 2(1):93–100, January 1991.

[204] C.C. Lee. Fuzzy logic in control systems: Fuzzy logic controller. *IEEE Transactions on Systems, Man, and Cybernetics*, 20(2):404–435, April 1990.

[205] J.X. Lee and G. Vukovich. The dynamic fuzzy logic system: Nonlinear system identification and application to robotic manipulators. *Journal of Robotic Systems*, 14(6):391–406, June 1997.

[206] M. Lee and H-S. Choi. A robust neural controller for underwter robot manipulators. *IEEE Transactions on Neural Networks*, 11(6):1465–1470, November 2000.

[207] M.L. Lee, H-S. Choi, and Y. Park. A robust neural controller for underwater robot manipulator. In *Proceedings of the IEEE International Joint Conference on Neural Networks*, pages 2098–2103, Anchorage, May 1998.

[208] S. Lee and R.M. Kil. Redundant arm kinematic control with recurrent loop. *Neural Networks*, 7(4):643–659, 1994.

[209] S. Lee and M.H. Kim. Learning expert systems for robot fine motion control. In *Proceedings of the IEEE International Symposium on Intelligent Control*, pages 534–544, Arlington, 1988.

[210] S. Lee and H.S. Lee. Generalized impedance of manipulators: Its application to force and position control. In *Proceedings of the 5th International Conference on Advanced Robotics*, pages 1477–1480, Pisa, 1991.

[211] F.L. Lewis, S. Jagannathan, and C.J. Harris. *Neural Network Control of Robot Manipulators and Nonlinear Systems*. Taylor & Francis, London, 1998.

[212] F.L. Lewis, K. Liu, and A. Yesildirek. Neural net based robot controller with quaranteed tracking performance. *IEEE Transactions on Neural Networks*, 6(3):703–715, 1995.

[213] M. Lewis, A. Fagg, and G.A. Bekey. Genetic algorithms for gait synthesis in a hexapod robots. In Y.Zheng, editor, *Recent Trends in Mobile Robots*, pages 317–331. World Scientific, Singapore, 1994.

[214] J-H. Li, P-M. Lee, and S-J. Lee. Neural net based nonlinear adaptive control for autonomous underwater vehicles. In *Proceedings of the IEEE International Conference on Robotcss and Automation*, pages 1075–1080, Washington, D.C., May 2002.

[215] W. Li, C. Ma, and F.M. Wahl. A neuro-fuzzy system architecture for behaviour-based control of a mobile robot in unknown environment. *Fuzzy Sets and Systems*, 87(2):133–140, 1997.

[216] Y.F. Li and C.C. Lau. Development of fuzzy algorithm for servo systems. *IEEE Control Systems Magazine*, 9(4):65–72, April 1989.

[217] C.M. Lim and T.I. Hiyama. Application of fuzzy logic control to a manipulator. *IEEE Transactions on Robotics and Automation*, 7(5):688–691, 1991.

[218] C-J. Lin and C-T. Lee. Reinforcement learning for an ART-based fuzzy adaptive learning control network. *IEEE Transactions on Neural Networks*, 7(3):709–731, May 1996.

[219] C-T. Lin and C.S.G. Lee. Neural-network-based fuzzy logic control and decision system. *IEEE Transactions on Computers*, 40(12):1320–1336, December 1991.

[220] C-T. Lin and C.S.G. Lee. Reinforcement structure / parameter learning for neural-network-based fuzzy logic control system. *IEEE Transactions on Fuzzy Systems*, 2(1):46–63, February 1994.

[221] L-R. Lin and H-P. Huang. Integrating fuzzy control of the dextrous NTU hand. *IEEE Transactions on Mechatronics*, 1(3):216–229, September 1996.

[222] H. Lipson and J.B. Pollack. Towards contionuously reconfigurable self-designing robotics. In *Proceedings of the IEEE International Conference on Robotics and Automation*, pages 1761–1766, San Francisko, April 2000.

[223] J. Lorentz and J. Yuh. A survey and experimental study of neural network AUV control. In *Proceedings of the IEEE AUV96*, Monterey, 1996.

[224] K.H. Low, W.K. Leow, and M.H. Ang.Jr. Integrated planning and control of mobile robot with self-organizing neural network. In *Proceedings of the IEEE International Conference on Robotcss and Automation*, pages 3870–3875, Washington, D.C., May 2002.

[225] P. Lugner. The influence of the structure of automobile models and tyre characteristics on the theoretical results of steady-state and transient vehicle performance. In *Proceedings of the 5th VSD-2nd IUTAM Symposium*, pages 21–39, 1984.

[226] H.H. Lund, J. Hallam, and W-P. Lee. Evolving robot morphology. In *Proceedings of the IEEE 4th International Conference on Evolutionary Computation*, 1997.

[227] R. Luo and T.M. Chen. Target tracking by grey prediction theory and look-ahead fuzzy logic control. In *Proceedings of the IEEE International Conference on Robotics and Automation*, pages 1176–1181, Detroit, May 1999.

[228] H. Maaref and C. Barret. Sensor-based fuzzy navigation of an autonomous mobile robot in an indoor environment. *Control Engineering Practice*, 8:757–768, 2000.

[229] R. Maas, V. Zahn, M. Dapper, and R. Eckmiller. Hard contact surface tracking for industrial manipulators woth (SR) position based force control. In *Proceedings of the IEEE International Conference on Robotics and Automation*, pages 1481–1486, Detroit, May 1999.

[230] H. Malki, D. Misir, D. Feigenspan, and G. Chen. Fuzzy PID control of a flexible-joint robot arm with uncertainties from time-varying loads. *IEEE Transactions on Control Systems Technology*, 5(3):371–378, May 1997.

[231] E.H. Mamdani. Application of fuzzy algorithms for control of simple dynamic plant. *IEE Proceedings Control and Science*, 121(12):1585–1588, December 1974.

[232] N.J. Mandic, E.M. Scharf, and E.H. Mamdani. Practical application of a heuristic fuzzy rule-based controller to the dynamic control of a robot arm. *IEE Proceedings, Part D*, 132:190–203, 1985.

[233] Z. Mao and T.C. Hsia. Obstacle avoidance inverse kinematics solution of redundant robots by neural network. *Robotica*, 15(1):3–10, 1997.

[234] D.W. Marhefka and D.E. Orin. Fuzzy control of quadrupedal running. In *Proceedings of the IEEE International Conference on Robotics and Automation*, pages 3063–3069, San Francisko, April 2000.

[235] T. Martinetz, H. Ritter, and K. Schulten. Three-dimensional neural net for learning visuo-motor coordination of a robot arm. *IEEE Transactions on Neural Networks*, 1(1):131–136, March 1990.

[236] E. Martinson, A. Stoychev, and R. Arkin. Robot behavioral selection using Q-learning. In *Proceedings of the IEEE/RSJ International Conference on Intelligent Robots and Systems*, pages 970–975, Lausanne, October 2002.

[237] M. Mataric and D. Cliff. Challenges in evolving controllers for physical robots. *Robot.Autonomous Syst.*, 19(1):67–83, 1996.

[238] M. Mataric and et al. Behaviour-based primitives for articulated control. In *Proceedings of the Fifth International Conference Soc.for Adaptive Behaviours*, pages 165–170, 1998.

[239] M.J. Mataric. *Interaction and Intelligent Behavior*. PhD thesis, MIT, Cambridge, 1994.

[240] G. Mauris, E. Benoit, and L. Foulloy. An intelligent ultrasonic range finding sensor for robotics. In *Proceedings of the 13th IFAC World Congress*, volume A, pages 487–492, San Francisko, 1996.

[241] R.V. Mayorga and P. Sanongboon. A radial basis function network approach for inverse kinematics and singularities prevention of redundant manipulators. In *Proceedings of the IEEE International Conference on Robotcss and Automation*, pages 1955–1960, Washington, D.C., May 2002.

[242] J.M. Mendel and K.S. Fu, editors. *Adaptive, Learning and Pattern Recognition Systems: Theory and Applications*. Academic Press, New York, 1970.

[243] G. Mester and S. Pletl. Genetic algorithm based structure optimization of the fuzzy control system. In *Proceedings of the 3rd ECPD International Conference on Advanced Robotics, Intrlligent Automation and Active Systems*, pages 209–214, Bremen, 1997.

[244] J.A. Meyer, P. Husbands, and I. Harvey. Evolutionary robotics: A survey of applications and problems. In P.Husbands and J.A.Meyer, editors, *Evolutionary Robotics*, volume 1468, pages 1–21. Springer-Verlag, Berlin,Germany, 1998.

[245] Z. Michalewicz. *Genetic Algorithms + Data Structures = Evolution Program*. Springer Verlag, Berlin, 1994.

[246] O. Miglino, H.H. Lund, and S. Nolfi. Evolving mobile robots in simulated and real environments. *Artificial Life*, 2:417–434, 1996.

[247] W. Miller, III. Real-time neural network control of a biped walking robot. *IEEE Control Systems Magazine*, pages 41–48, February 1994.

[248] W. Miller, III, F. Glanz, and L. Kraft. Application of a general learning algorithm to the control of robotic manipulators. *International Journal of Robotics Reserach*, 6(2), 1987.

[249] W. Miller III. Sensor-based control of manipulation robots using a general learning algorithm. *IEEE Journal on Robotics and Automation*, 3:157–165, April 1987.

[250] W. Miller III. Real-time application of neural networks for sensor-based control of robots with vision. *IEEE Transactions on Systems,Man, and Cybernetics*, 19(4):825–831, 1989.

[251] M. Mizumoto. Fuzzy controls under various approximate reasoning methods. In *Proceedings of the 2nd IFSA Congress*, pages 143–146, Tokyo, 1987.

[252] M.B. Montaner and A. Ramirez-Serrano. Fuzzy knowledge-based controller design for autonomous robot navigation. *Expert Systems with Applications*, 14:179–186, 1998.

[253] J.F. Montgomery. Learning helicopter control through teaching by showing. Technical Report IRIS-99-370, University of Southern California, Los Angeles, 1999.

[254] A. Moran, T. Hasegawa, and M. Nagai. Continuously controlled semi-active suspension using neural networks. In *Proceedings of International Symposium on Advanced Vehicle Control*, pages 305–310, Tsukuba, 1994.

[255] S. Muarakami, F. Takemoto, H. Fujumura, and E. Ide. Weld-line tracking control of arc welding robot using fuzzy logic controller. *Fuzzy Sets and Systems*, 32(4):221–237, 1989.

[256] F.A. Mussa-Ivaldi and E. Bizzi. Learning newtonian mechanics. In P. Morasso and V. Sanguineti, editors, *Self-organization, Computational Maps and Motor Control*, pages 491–501. Elsevier, Amsterdam, 1997.

[257] M. Nagai, E. Ueda, and A. Moran. Integration of linear systems and neural networks for the design of nonlinear four-wheel-steering systems. In *Proceedings of International Symposium on Advanced Vehicle Control*, pages 153–158, Tsukuba, 1994.

[258] K. Nagasaka, A. Konno, M. Inaba, and H. Inoue. Acquisition of visually guided swing motion based on genetic algorithms and neural networks in two-armed bipedal robot. In

Proceedings of the IEEE International Conference on Robotics and Automation, pages 2944–2949, Albuquerque, April 1997.

[259] K.S. Narendra and K. Parthasarathy. Identification and control of dynamical systems using neural networks. *IEEE Transactions on Neural Networks*, 1(1):4–27, 1990.

[260] O. Nerrand, P. Roussel-Ragot, D. Urbani, L. Personnaz, and G. Dreyfus. Training recurrent neural networks: Why and how ? An illustration in dynamical process modeling. *IEEE Transactions on Neural Networks*, pages 178–184, March 1994.

[261] K.C. Ng and M.M. Trivedi. A neuro-fuzzy controller for mobile robot navigation and multirobot convoying. *IEEE Transactions on Systems, Man, and Cybernetics, Part B:Cybernetics*, 28(6):829–840, December 1998.

[262] D. Nikovski and I. Nourbakhsh. Learning probabilistic models for state tracking of mobile robotslifelong evolution for adaptive robots. In *Proceedings of the IEEE/RSJ International Conference on Intelligent Robots and Systems*, pages 1026–1031, Lausanne, October 2002.

[263] S. Nolfi. Evolving nontrivial behaviour in autonomous robots: Adaptation is more powerful than decomposition and integration. In T.Gomi, editor, *Evolutionary Robotics: From Intelligent Robotics to Artificial Life*. AAI Books, Kanata,ON,Canada, 1997.

[264] S. Nolfi and D. Floreano. *Evolutionary Robotics: Biology, Intelligence and Technology of Self-Organizing Machines*. The MIT Press, Cambridge, 2001.

[265] L. Ojeda and J. Borenstein. FLEXnav: Fuzzy logic expert rule-based position. In *Proceedings of the IEEE International Conference on Robotcss and Automation*, pages 317–322, Washington, D.C., May 2002.

[266] E. Ono, M. Yamashita, and Y. Yamaguchi. Distributed hierarchy control of active suspension system. In *Proceedings of International Symposium on Advanced Vehicle Control*, pages 373–378, Tsukuba, 1994.

[267] E. Oyama and S. Tachi. Modular neural net system for inverse kinematic learning. In *Proceedings of the IEEE International Conference on Robotics and Automation*, pages 3239–3246, San Francisko, April 2000.

[268] T. Ozaki, T. Suzuki, T. Furuhashi, S. Okuma, and Y. Uchikawa. Trajectory control of robotic manipulators using neural networks. *IEEE Transactions on Industrial Electronics*, 38(3):195–202, June 1991.

[269] C. Pal, I. Hagiwara, S. Morishita, and H. Inoue. Application of neural networks in real time identification of dynamic structural response and prediction of road - friction coefficients from steady state automobile response. In *Proceedings of International Symposium on Advanced Vehicle Control*, pages 527–632, Tsukuba, 1994.

[270] R. Palm. Fuzzy controller for a sensor guided robot manipulator. *Fuzzy Sets and Systems*, 31(2):133–149, 1989.

[271] H.D. Patino, R. Carelli, and B.R. Kuchen. Neural networks for advanced control of robot manipulators. *IEEE Transactions on Neural Networks*, 13(2):343–354, March 2002.

[272] H-L. Pei, Q-J. Zhou, and T.P. Leung. A neural network robot force controller. In *Proceedings of the IEEE/RSJ International Conference on Intelligent Robots and Systems*, pages 1974–1979, Raleigh, July 1992.

[273] H. Peng and M. Tomizuka. Program on advanced technology (PATH) program - lateral control of front-wheel-steering rubber-tire vehicles. Technical Report UCB-ITS-PRR-90-5, Institute of Transportation Studies, University of California at Berkeley, 1990.

[274] J. Pettersson, H. Sandholt, and H. Wahde. A flexible evolutionary method for the generation and implementation of behaviours for humanoid robots. In *Proceedings of the IEEE-RAS International Conference on Humanoid Robots HUMANOIDS 2001*, Tokyo, November 2001.

[275] H. Pham, K. Hedrick, and M. Tomizuka. Autonomous steering and cruise control of automobiles via sliding mode control. In *Proceedings of International Symposium on Advanced Vehicle Control*, pages 444–448, Tsukuba, 1994.

[276] J.B. Pollack, H. Lipson, S. Ficici, P. Funes, G. Hornby, and R. Watson. Evolutionary techniques in physical robotics. In A. Thompson, J. Miller, T. C. Fogarty, and P. Thomson, editors, *Proceedings of the Third International Conference on Evolvable Systems: from Biology to Hardware*. Springer Verlag, Berlin, 2000.

[277] D. Popovic and R.S. Shekhawat. A fuzzy expert tuner for robot controller. In *Proceedings of the IFAC Symposium on Robot Control*, pages 229–233, Vienna, 1991.

[278] F. Pourboghrat. Neural networks for learning inverse kinematics of redundant manipulators. In *Proceedings of the 32nd Midwest Symposium on Circuits Systems*, pages 760–762, 1990.

[279] T.J. Procyk and E.H. Mamdani. A linguistic self-organizing process controller. *Automatica*, 15(1):15–30, January 1979.

[280] D.R. Ramirez, D. Limon, J. Gomez-Ortega, and E.F. Camacho. Nonlinear MBPC for mobile robot navigation using genetic algorithms. In *Proceedings of the IEEE International Conference on Robotics and Automation*, pages 2452–2457, Detroit, May 1998.

[281] T. Reil and P. Husbands. Evolution of central pattern generators for bipedal walking in a real-time physics environment. *IEEE Transactions on Evolutionary Computation*, 6(2):159–168, April 2002.

[282] O.R. Rioul and M. Veterli. Wavelets and signal processing. *IEEE Signal Processing Magazine*, 8:14–38, 1991.

[283] A.D. Rodic and M.K. Vukobratovic. Contribution to the integrated control synthesis of road vehicles. *IEEE Transactions on Control Systems Technology*, 7(1):64–78, 1999.

[284] M. Rucci and P. Dario. Autonomous learning of tactile-motor coordination in robotics. In *Proceedings of the IEEE International Conference on Robotics and Automation*, pages 3230–3236, San Diego, May 1994.

[285] D.E. Rumelhart and J.L. McClelland. *Parallel Distributed Processing (PDP): Exploration in the Microstructure of Cognition*, volume 1-2. MIT Press, Cambridge, 1986.

[286] A. Saffiotti, E. Ruspini, and K. Konolige. Robust execution of robot plans using fuzzy logic. In *Proceedings of the Workshop on Fuzzy Logic in Artificial Intelligence*, pages 24–37, Chambery, 1993.

[287] G. Sahar and J.M. Hollerbach. Planning of minimmum-time trajectory for robot arms. *International Journal of Robotics Research*, 5(3):91–100, 1986.

[288] W. Salatian, K. Yi, and Y. Zheng. Reinforcement learning for a biped robot to climb sloping surfaces. *Journal of Robotic Systems*, pages 283–296, April 1997.

[289] W. Salatian and Y.F. Zheng. Gait synthesis for a biped roboti climbing sloping surfaces using neural networks - I. Static learning. In *Proceedings of the IEEE International Conference on Robotics and Automation*, pages 2601–2606, Nice, May 1992.

[290] W. Salatian and Y.F. Zheng. Gait synthesis for a biped roboti climbing sloping surfaces using neural networks - II. Dynamic learning. In *Proceedings of the IEEE International Conference on Robotics and Automation*, pages 2607–2611, Nice, May 1992.

[291] R.M. Sanner and J.J.E. Slotine. Stable adaptive control of robot manipulators using neural networks. *Neural Computation*, 7(3):753–790, 1995.

[292] V. Santibanez, R. Kelly, and M.A. Lliama. Fuzzy PD+ control for robot manipulators. In *Proceedings of the IEEE International Conference on Robotics and Automation*, pages 2112–2117, San Francisko, April 2000.

[293] G.N. Saridis. Knowledge implementation: Structures of intelligent control systems. In *Proceedings of the IEEE International Symposium on Intelligent Control*, pages 9–17, Philadelphia, January 1987.

[294] S. Schaal. Advances in neural information processing systems 9. In M. C. Mozer, M. Jordan, and T. Petsche, editors, *Self-organization, Computational Maps and Motor Control*, pages 1040–1046. MIT Press, Cambridge, 1997.

[295] S. Schaal. Is imitation learning the route to humanoid robots? *Trends in Cognitive Sciences*, 3:233–242, 1999.

[296] S. Schaal and C.G. Atkeson. Constructive incremental learning from only local information. *Neural Computation*, 10:2047–2084, 1998.

[297] S. Schaal and D. Sternad. Programmable pattern generators. In *Proceedings of the 3rd International Conference on Computational Intelligence in Neuroscience*, pages 48–51, Research Triangle Park, NC, 1998.

[298] N. Mandic Scharf, E.M. and E. Mamdani. A self-organizing algorithm for the control of a robot arm. *International Journal of Robotics and Automation*, 1(1):33–41, 1986.

[299] H. Seraji and A. Howard. Behavior-based robot navigation on challenging terrain: A fuzzy logic approach. *IEEE Transactions on Robotics and Automation*, 18(3):308–321, June 2002.

[300] D. Seslija and M. Vukobratovic. Environment parameters identification for the control of robotized machining. In *Proceedings of the first ECPD International Conference on Advanced Robotics and Intelligent Automation*, pages 632–637, Athens, September 1995.

[301] K. Seung-Woo and P. Mignon. Fuzzy compliance robot control. In *Proceedings of the IEEE Symposium on Intelligent Robots and Systems*, pages 1628–1631, 1991.

[302] R. Sharma and N. Srinivasa. A framework for robot control with active vision using a neurel network based spatial representation. In *Proceedings of the IEEE International Conference on Robotics and Automation*, pages 1966–1971, Minneapolis, April 1996.

[303] M. Shibata, T. Murakami, and K. Ohnishi. A unified approach to position and force control by fuzzy logic. *IEEE Transactions on Industrial Electronics*, 43(1):81–87, February 1996.

[304] T. Shibata and T. Fukuda. Hierarchical intelligent control for robotic motion. *IEEE Transactions on Neural Networks*, 5(5):823–832, September 1994.

[305] S.E. Shladover, C.A. Desoer, J.K. Hedrick, M. Tomizuka, J. Arland, W. Zhang, D.H. McMahon, H. Peng, S. Sheikholeslam, and N. McKeown. Automatic vehicle control developments in the PATH program. *IEEE Transactions on Vehicular Technology*, 40(1):114–130, 1991.

[306] E.D.V. Simoes and D.A.C. Barone. Predation: an approach to improving the evolution of real robots with a distributed evolutionary controller. In *Proceedings of the IEEE International Conference on Robotics and Automation*, pages 664–669, Washington, D.C., 2002.

[307] K. Sims. Evolving 3d morphology and behavior by competition. In R. Brooks and P. Maes, editors, *Proceedings of the Fourth Workshop on Artificial Life*, pages 28–39. MIT Press, Boston, 1994.

[308] M. Skubic and R. Volz. Identifying contact formation from sensory patterns and its application to robot programming in demonstration. In *Proceedings of the IEEE/RSJ International Conference on Intelligent Robots and Systems*, pages 458–464, Osaka, November 1996.

[309] M. Skubic and R. Volz. Learning force sensory patterns and skills from human demonstration. In *Proceedings of the IEEE International Conference Robotics and Automation*, pages 284–290, Albuquerque, April 1997.

[310] M. Skubic and R. Volz. Acquiring robust, force-based assembly skills from human demonstration. *IEEE Transactions on Robotics and Automation*, 16(6):772–781, December 2000.

[311] M. Skubic and R. Volz. Identifying sigle-ended contact formations from force sensor patterns. *IEEE Transactions on Robotics and Automation*, 16(5):597–602, October 2000.

[312] C. Son. Optimal planning technique with a fuzzy co-ordinator for an intelligent robot's part assembly. *IEE Proceedings Control Theory and Applications*, 144(1):45–52, January 1997.

[313] K-T. Song and W-Y. Sun. Robot control optimization using reinforcement learning. *Journal of Intelligent & Robotic Systems*, 21(3):221–238, March 1998.

[314] Y.D. Song. Neuro-adaptive control with application to robotic systems. *Journal of Robotic Systems*, 14(6):433–448, June 1997.

[315] B. Srinivasan, U.R. Prasad, and N.J. Rao. Back propagation through adjoints for the identification of nonlinear dynamic systems using recurrent neural models. *IEEE Transactions on Neural Networks*, pages 213–228, March 1994.

[316] M.S. Stachowicz and M.E. Kochanska. Fuzzy modeling of the process. In *Proceedings of the 2nd IFSA Congress*, pages 86–89, Tokyo, 1987.

[317] K. Stanley, Q.M.J. Wu, A. Jerbi, and W.A. Gruver. Neural network-based vision guided robotics. In *Proceedings of the IEEE International Conference on Robotics and Automation*, pages 281–286, Detroit, May 1999.

[318] K.M. Stellakis and K.P. Valavanis. Fuzzy logic-based formulation of the organizer of intelligent robotic systems. *Journal of Intelligent and Robotic Systems*, 4(1):1–24, 1991.

[319] D. Stipanicev and J. Efstathiou. Fuzzy reasoning in planning, decision making and control: Intelligent robots,vision,natural language. In B.Sou£ek, editor, *Fuzzy,Holographic and Parallel Intelligence*, pages 93–132. John Wiley & Sons,Inc., New York, 1992.

[320] R. Stojic and O. Timcenko. A two-level hybrid control for bipedal walking. In *Proceedings of the 5th IFAC Symposium on Robot Control*, pages 801–806, Nantes, 1997.

[321] M. Sugeno and K. Murakami. An experimental study on fuzzy parking control using a model car. In M.Sugeno, editor, *Industrial Application of Fuzzy Control*, pages 125–138. North-Holland, 1985.

[322] M. Sugeno and M. Nishida. Fuzzy control of model car. *Fuzzy Sets and Systems*, 16:103–113, 1985.

[323] F. Sun, Z. Sun, and P-Y. Woo. Neural network-based adaptive controller design of robotic manipulators with an observer. *IEEE Transactons on Neural Networks*, 12(1):54–67, January 2001.

[324] L. Sun and C. Doeschner. Visuo-motor coordination of a robot manipulator based on neural networks. In *Proceedings of the IEEE International Conference on Robotics and Automation*, pages 1737–1742, Leuven, May 1998.

[325] R.S. Sutton and A.G. Barto. *Reinforcement Learning: An Introduction*. MIT Press, Cambridge, 1998.

[326] R.S. Sutton, D. Precup, and S.N. Singh. Between mdps and semi-mdps:A framework for temporal abstraction in reinforcement learning. *Artificial Intelligence*, 112:181–211, 1999.

[327] D. Suwimonteerabuth and P. Chongstitvatana. On-line robot learning by reward and punishment for a mobile robot. In *Proceedings of the IEEE/RSJ International Conference on Intelligent Robots and Systems*, pages 921–926, Lausanne, October 2002.

[328] R. Syam, K. Watanabe, K. Izumi, and K. Kiguchi. Control of nonholonomic mobile robot by an adaptive actor-critic method with simulated experience based value-functions. In *Proceedings of the IEEE International Conference on Robotcss and Automation*, pages 3960–3965, Washington, D.C., May 2002.

[329] H.H. Szu, B. Telfer, and S. Kadambe. Neural network adaptive wavelets for signal representation and classification. *Optical Engineering*, 31(9):1907–1916, September 1992.

[330] T. Takagi and M. Sugeno. Derivation of fuzzy control rulles from human operator's control actions. In *Proceedings of the IFAC Symposium on Fuzzy Information, Knowledge Representation and Decision Analysis*, pages 55–60, July 1983.

[331] T. Takagi and M. Sugeno. Fuzzy identification of systems and its applications to modeling and control. *IEEE Transactions on Systems, Man, and Cybernetics*, 15(1):116–132, January 1985.

[332] T. Takeuchi. An autonomous fuzzy mobile robot. *Advanced Robotics*, 5(2):215–230, 1991.

[333] K. Tanaka, M. Shimizu, and K. Tsuchiya. A solution to an inverse kinematics problem of a redundant manipulator using neural networks. In K.Kohonen, editor, *Artificial Neural Networks*. Elsevier, 1991.

[334] W-S. Tang and J. Wang. Two recurrent neural network for local joint torque optimisation of kinematically redundant manipulators. *IEEE Transactions on Systems,Man, and Cybernetics, Part B*, 30:120–128, February 2000.

[335] W-S. Tang and J. Wang. A recurrent neural network for minimum infinity-norm kinematic control of redundant manipulators with an improved problem formulation and reduced architecture complexity. *IEEE Transactions on Systems,Man, and Cybernetics, Part B*, 31:98–105, February 2001.

[336] R. Tanscheit and E.M. Scharf. Experiments with the use of a rule-based self-organizing controller for robotics applications. *Fuzzy Sets and Systems*, 26:195–214, 1988.

[337] J.M. Tao. Application of neural network with real-time training to robust position/force control. In *Proceedings of the IEEE International Conference on Robotics and Automation*, pages 124–148, Atlanta, May 1993.

[338] J.M. Tao and J.Y.S. Luh. Application of neural network with real-time training to robust position/force control of multiple robots. In *Proceedings of the IEEE International Conference on Robotics and Automation*, pages 142–148, Atlanta, May 1993.

[339] M. Tarokh and S. Bailey. Force tracking with unknown environment parameters using adaptive fuzzy controllers. In *Proceedings of the IEEE International Conference on Robotics and Automation*, pages 270–275, Minneapolis, April 1996.

[340] M. Tarokh and S. Bailey. Adaptive fuzzy force control of manipulators with unknown environment parameters. *Journal of Robotic Systems*, 14(5):341–353, May 1997.

[341] T. Terano, K. Asai, and M. Sugeno. *Fuzzy Systems Theory and Its Applications*. Academic Press, Boston, 1992.

[342] A. Thompson. Evolving electronic robot controllers that exploit hardware resources. In F. et al. Moran, editor, *Advances in Artificial Life: Proc. of ECAL95*. Springer Verlag, Barselona, 1995.

[343] O. Timcenko and R. Stojic. On the control of bipedal walking subject to lateral disturbances. In *Proceedings of the 2nd ECPD International Conference on Advanced Robotics,Intelligent Automation and Active Systems*, pages 282–287, Vienna, 1996.

[344] L.H. Tsoukalas, E.N. Houstis, and G.V. Jones. Neurofuzzy motion planners for intelligent robots. *Journal of Intelligent and Robotic Systems*, 19:339–356, 1997.

[345] S. Turk and M. Otter. The DFVLR Models 1 and 2 of the Manutec r3 robot. *Robotersysteme*, (3):753–768, 1987. (in German).

[346] C. Tzafestas and S. Tzafestas. Recent algorithms for fuzzy and neurofuzzy path planning and navigation of autonomous mobile robots. In *Proceedings of the European Control Conference*, Karlsrue, September 1999.

[347] S. Tzafestas. Fuzzy and neural approaches to robot control. In *Proceedings of the 1st ECPD International Conference, Advanced Robotics and Intelligent Automation*, pages 34–55, Athens, September 1995.

[348] S. Tzafestas and N.P. Papanikolopoulos. Incremental fuzzy expert PID control. *IEEE Transactions on Industrial Electronics*, 37:365–371, October 1990.

[349] S. Tzafestas and K.C. Zikides. A 3-level neuro-fuzzy autonomous robot navigation system. In *Proceedings of the European Control Conference*, Brussels, June 1997.

[350] G.J. Vachtsevanos, K. Davey, and K.M. Lee. Development of novel intelligent robotic manipulator. *IEEE Control Systems Magazine*, 7(3):9–15, 1987.

[351] K.P. Valavanis and G.N. Saridis, editors. *Intelligent Robotic Systems*. Kluwer Academic Publishers, Norwell, 1992.

[352] S.T. Venkataraman, S. Gullati, J. Barhen, and N. Toomarian. A neural network based identification of environment models for compliant control of space robots. *IEEE Transactions on Robotics and Automation*, 9(5):685–697, October 1993.

[353] S. Vijayakumar and S. Schaal. Fast and efficient incremental learning for high-dimensional movement systems. In *Proceedings of the International Conference on Conference on Robotics and Automation*, San Francisko, 2000.

[354] S. Vijayakumar and S. Schaal. Real time learning in humanoids: A challenge for scalability algorithms. In *Proceedings of the IEEE-RAS International Conference on Humanoid Robots HUMANOIDS 2000*, Cambridge, September 2000.

[355] M. Vukobratovic and Y. Ekalo. Unified approach to control laws synthesis for robotic manipulators in contact with dynamic environment. In *Tutorial S5: Force and Contact Control in robotic systems, IEEE International Conference on Robotics and Automation*, pages 213–229, Atlanta, May 1993.

[356] M. Vukobratovic and Y. Ekalo. New approach to the control of manipulation robots interacting with dynamic environment. *Robotica*, 14:31–39, 1996.

[357] M. Vukobratovic and B. Karan. Experiments with fuzzy logic robot control with model-based dynamic compensation. In *Proceedings of the 25th International Symposium on Industrial Robots*, pages 215–222, Hannover, 1994.

[358] M. Vukobratovic and B. Karan. Experiments with fuzzy logic robot control with model-based dynamic compensation in nonadaptive decentralized control scheme. *IASTED International Journal of Robotics and Automation*, 11(3):118–131, 1996.

[359] M. Vukobratovic and D. Katic. Connectionist control structures for high-efficiency learning in robotics. In S.Tzafestas, editor, *Applied Control*, pages 705–753. Marcel Dekker, New York, 1993.

[360] M. Vukobratovic and D. Katic. Connectionist learning control algorithms for contact tasks in industrial robotics. In *Proceedings of the 24th International Symposium on Industrial Robots*, Tokyo, November 1993.

[361] M. Vukobratovic, D. Stokic, and N. Kircanski. *Non-Adaptive and Adaptive Control of Manipulation Robots*. Springer Verlag, Berlin, 1985.

[362] M. Vukobratovic and D. Surdilovic. Control of robotic systems in contact tasks, An overiew. In *Tutorial S5: Force and Contact Control in Robotic Systems: IEEE International Conference on Robotics and Automation*, pages 13–32, Atlanta, May 1993.

[363] M. Vukobratovic and O. Timcenko. Stability analysis of certain class of bipedal walking robots with hybridization of classical and fuzzy control. In *Proceedings of the 1st ECPD International Conference on Advanced Robotics and Intelligent Automation*, pages 290–295, Athens, 1995.

[364] B.A.M. Wakileh and K.F. Gill. Use of fuzzy logic in robotics. *Computers in Industry*, 10:33–46, 1988.

[365] J. Walker and M. Wilson. Lifelong evolution for adaptive robots. In *Proceedings of the IEEE/RSJ International Conference on Intelligent Robots and Systems*, pages 984–989, Lausanne, October 2002.

[366] H. Wang, T.T. Lee, and W.A. Gruver. A neuromorphic controller for a three link biped robot. *IEEE Transactions on Systems, Man, and Cybernetics*, 22(1):164–169, January/February 1992.

[367] J. Wang, Q. Hu, and D. Jiang. A lagrange network for kinematic control of redundant robot manipulators. *IEEE Transactions on Neural Networks*, 10:1123–1132, 1999.

[368] J-S. Wang and C.S.G. Lee. An on-line self-organizing neuro-fuzzy control for autonomous underwater vehicles. In *Proceedings of the IEEE International Conference on Robotics and Automation*, pages 2416–2421, Detroit, May 1999.

[369] J-S. Wang and C.S.G. Lee. Self-adaptive recurrent neuro-fuzzy control for an autonomous underwater vehicle. In *Proceedings of the IEEE International Conference on Robotcss and Automation*, pages 1095–1100, Washington, D.C., May 2002.

[370] J-S. Wang, C.S.G. Lee, and J. Yuh. Self-adaptive neuro-fuzzy with fast parameter learning for autonomous underwater vehicle control. In *Proceedings of the IEEE International Conference on Robotics and Automation*, pages 3861–3866, San Francisko, April 2000.

[371] L-X. Wang. *Adaptive Fuzzy Systems and Control:Design and Stability Analysis*. Prentice Hall, Englewood Cliffs, 1994.

[372] Q. Wang and A.M.S. Zalzala. Investigations into robotic multi-joint motion considering multi-criteria optimisation using genetic algorithms. In *Proceedings of the 13th IFAC World Congress*, volume A, pages 301–306, San Francisko, 1996.

[373] Y. Wang, B. Thibodeau, A.H. Fagg, and R. Grupen. Learning optimal switching policies for path tracking tasks on a mobile robot. In *Proceedings of the IEEE/RSJ International Conference on Intelligent Robots and Systems*, pages 915–920, Lausanne, October 2002.

[374] K. Ward and A. Zelinsky. Acquiring mobile robot behaviour by learning trajectory velocities with multiple FAM matrices. In *Proceedings of the IEEE International Conference on Robotics and Automation*, pages 668–673, Leuven, May 1998.

[375] P.D. Wasserman. *Neural Computing - Theory and Practice*. Van Nostrand Reinhold, New York, 1989.

[376] K. Watanabe, M.M.A. Hashem, and K. Izumi. Global path planning of mobile robots as an evolutionary control problem. In *Proceedings of the European Control Conference*, Karlsrue, September 1999.

[377] K. Watanabe, J. Tang, M. Nakamura, S. Koga, and T. Fukuda. A fuzzy-gaussian neural network and its application to mobile robot control. *IEEE Transactions on Control Systems Technology*, 4(2):193–199, March 1996.

[378] C.J.C.H. Watkins. *Learning from Delayed Rewards*. PhD thesis, King's College, Cambridge, 1989.

[379] R. Watson, S. Ficici, and J.B. Pollack. Embodied evolution: Embodying an evolutionary algorithm in a population of robots. In P. Angeline, M.Michaelewicz, G. Schonauer, X. Yao, and Z. Zalzala, editors, *Proceedings of the 1999 Congress on Evolutionary Computation*. 1999.

[380] D.E. Whitney. Historical perspective and state of the art in robot force control. *International Journal of Robotics Research*, 6(1):3–14, 1987.

[381] B. Widrow. Generalization and information storage in networks of Adaline neurons. In *Self-Organizing Systems*, pages 435–461. Spartan Books, 1962.

[382] C-J. Wu. A learning fuzzy algorithm for motion planning of mobile robots. *Journal of Intelligent and Robotic Systems*, 11:209–221, 1995.

[383] Y. Xia and J. Wang. A dual neural network for kinematic control of redundant robot manipulators. *IEEE Transactions on Systems,Man, and Cybernetics, Part B*, 31:147–154, February 2001.

[384] J. Xiao, Z. Michalewicz, L. Zhang, and K. Trojanowski. Adaptive evolutionary planner/navigator for mobile robots. *IEEE Transactions on Evolutionary Computation*, 1(1):18–28, April 1997.

[385] M. Xiaowei, Y. Fu, W. Wei, M. Yulin, and C. Hegao. The path planning of mobile manipulator with genetic-fuzzy controller in flexible manufacturing cell. In *Proceedings of the IEEE International Conference on Robotics and Automation*, pages 329–334, Detroit, May 1999.

[386] C.W. Xu and Y.Z. Lu. Fuzzy model identification and self-learning for dynamic systems. *IEEE Transactions on Systems, Man, and Cybernetics*, 17(4):683–689, August 1987.

[387] Y. Xu and M. Nechyba. Fuzzy inverse kinematic mapping: Rule generation, efficiency and implementation. Technical Report CMU-RI-TR-93-02, Carnegie Mellon University, Pittsburgh, 1993.

[388]   A. Yang and K.H. Low. Fuzzy position/force control of a robot leg with a flexible gear system. In *Proceedings of the IEEE International Conference on Robotics and Automation*, pages 2159–2164, Washington, D.C., May 2002.

[389]   B-H. Yang and H. Asada. Hybrid linguistic / numeric control of deburring robots. In *Proceedings of the IEEE International Conference on Robotics and Automation*, pages 1467–1474, Nice, May 1992.

[390]   J-M. Yang, J-T. Horng, and C-Y. Kao. A new evolutionary approach to developing neural autonomous agents. In *Proceedings of the IEEE International Conference on Robotics and Automation*, pages 1411–1416, Detroit, May 1999.

[391]   S. Yasunobu and N. Minamiyama. A proposal of intelligent vehicle control system by predictive fuzzy control with hierarchical temporary target setting. In *Proceedings of the Conference Fuzz-IEEE '96*, pages 873–878, New Orleans, 1996.

[392]   D-Y. Yeung and G.A. Bekey. Using a contex-sensitive learning network for robot arm control. In *Proceedings of the IEEE International Conference on Robotics and Automation*, pages 1441–1447, Scottsdale, May 1989.

[393]   K-Y. Young and S-J. Shiah. An approach to enlarge learning space coverage for robot learning control. *IEEE Transactions on Fuzzy Systems*, 5(4):511–522, November 1997.

[394]   Z. Yu, H. Chen, and P. Woo. Fuzzy continuous gain scheduling H-infinity control based on taylor series fitting for robotic manipulators. In *Proceedings of the IEEE/RSJ International Conference on Intelligent Robots and Systems*, pages 2175–2080, Lausanne, October 2002.

[395]   J. Yuh. A neural net controller for underwater robotic vehicles. *IEEE Journal of Oceanic Engineering*, 15(3):161–166, 1990.

[396]   J. Yuh. Learning control for underwater robotic vehicles. *IEEE Control Systems Magazine*, 14(2):39–46, 1994.

[397]   L.A. Zadeh. Fuzzy algorithms. *Information and Control*, 12:94–102, 1968.

[398]   L.A. Zadeh. Quantitative fuzzy semantics. *Information Sciences*, 3:159–176, 1971.

[399]   L.A. Zadeh. A rationale for fuzzy control. *Journal of Dynamic Systems, Measurement and Control*, 94:3–4, 1972.

[400]   L.A. Zadeh. Outline of a new approach to the analysis of complex systems and decision processes. *IEEE Transactions on Systems, Man, and Cybernetics*, 3(1):28–44, January 1973.

[401]   L.A. Zadeh. The concept of a linguistic variable and its application to approximate reasoning. *Information Sciences*, 8:199–249,301–357, 1975.

[402]   L.A. Zadeh. Fuzzy logic. *IEEE Computer*, 21(4):83–93, April 1988.

[403]   L.A. Zadeh. Fuzzy sets and systems. *Information and Control*, 8:338–353, 1992.

[404]   V. Zahn, R. Maas, M. Dapper, and R. Eckmiller. Neural force control (NFC) applied to industrial manipulators in interaction with moving rigid objects. In *Proceedings of the*

IEEE International Conference on Robotics and Automation, pages 2780–2785, Detroit, May 1999.

[405] J. Zhang, R. Schmidt, and A. Knoll. Appearance-based visual learning in a neuro-fuzzy model for fine-positioning of manipulators. In *Proceedings of the IEEE International Conference on Robotics and Automation*, pages 1164–1169, Detroit, May 1999.

[406] M. Zhang, S. Peng, and Q. Meng. Neural network and fuzzy logic techniques based collision avoidance for a mobile robot. *Robotica*, 15(6):627–632, November-December 1997.

[407] Y. Zhang, J. Wang, and Y. Xu. A dual neural network for bi-criteria kinematic control of redundant manipulators. *IEEE Transactions on Robotics and Automation*, 18(6):923–931, December 2002.

[408] M. Zhao, N. Ansari, and E. Hou. Mobile manipulator path planning by a genetic algorithm. In *Proceedings of the IEEE/RSJ International Conference on Intelligent Robots and Systems*, pages 681–688, Raleigh, July 1992.

[409] C. Zhou and Q. Meng. Reinforcement learning with fuzzy evaluative feedback for a biped robot. In *Proceedings of the IEEE International Conference on Robotics and Automation*, pages 3829–3834, San Francisko, April 2000.

[410] H-J. Zimmermann. *Fuzzy Sets Theory - and its Applications*. Kluwer Academic, 1990.

Index

About the Authors

Dusko Katic was born in Kragujevac, Serbia, 1959. He received the B.S and M.S degrees in mechanical engineering from the University of Belgrade, Yugoslavia, in 1982 and 1987, and Ph.D. degree in electrical engineering from the University of Belgrade, Yugoslavia in 1994. Since 1983, he has been a researcher with the Robotics Laboratory, Mihajlo Pupin Institute, Belgrade, Yugoslavia, where he is involved in the modeling and design of advanced control systems for manipulation robots and other large-scale dynamic systems. From 1984 he was temporary teaching assistant and lecturer at the many universities at the former Yugoslavia. He is the author of about 80 scientific papers n the field of robotics in many leading international journals and conferences. His current scientific interests include theoretical and experimental work in the areas of intelligent autonomous systems, intelligent control, robotics, adaptive and learning systems, neural networks, evolutionary computing and signal processing.

Miomir Vukobratovic was born in Botos, Serbia, 1931. He received the B.Sc. and his first Ph.D. degrees in Mechanical Engineering from the University of Belgrade in 1957 and 1964 respectively, and his the second D.Sc. degree from the Institute of Mashinovedenya, (Central Institute in Mechanical Engineering), Soviet (now Russian) Academy of Sciences, Moscow,

1972. Since 1968 he was head of Biodynamics Department, then director of the Laboratory for Robotics and Flexible Automation and director of the Robotics Center, respectively at Mihailo Pupin Institute. He is a founder and scientific leader of the Belgrade school of Robotics. His major interest is in the development of efficient computer-aided modeling of robotic systems dynamics. His special interest being dynamic non-adaptive and adaptive control of non-contact and contact tasks in manipulation robotics, dynamic modeling stability, and control in biped locomotion, as well as intelligent control. He is author or coauthor of numerous scientific papers in the field of robotics and system theory, published in leading international journals. He is first author of 12 research monographs published in English, Japanese, Russian, Chinese and Serbian, advanced textbooks in robotics, and several chapters in international monographs. M. Vukobratovic is a full member of the Serbian Academy of Sciences and Arts, foreign member of the Soviet (now Russian) Academy of Sciences, full member of International Academy of Engineering, full member of the International Academy of Nonlinear Sciences, and other international academies. He is first president of the Yugoslav Academy of Engineering. He is Doctor honoris causa of the Moscow State University named after M. V. Lomonosov, and other European universities.

International Series on
MICROPROCESSOR-BASED AND INTELLIGENT SYSTEMS ENGINEERING

Editor: Professor S. G. Tzafestas, *National Technical University, Athens, Greece*

International Series on
MICROPROCESSOR-BASED AND
INTELLIGENT SYSTEMS ENGINEERING

KLUWER ACADEMIC PUBLISHERS – DORDRECHT / BOSTON / LONDON